INNOVATION & BALANCE
RESIDENT ARCHITECTURE

Architectures with Energy Saving, Low Carbon, Environmental Materials, Technology and Reuse Ideas

创新与平衡·住宅建筑
综合商住·别墅·公寓

江苏科学技术出版社

D. Schulz

Daniel Schulz
Architect - M.A. M.Arch. Dipl.-Ing.

Daniel Schulz, born in 1977 in Germany, was educated in the field of monument protection, sustainable and ecological design, possessing the German Diploma of Engineer in Architecture and later the International Master of Art in Architecture at the University of Wismar. In 2004, he continued and got a postgraduate Master of Architecture in China. After having design experiences in Germany and Australia, he went back to China as a principal architect for the Australian company, Jackson Architecture, at the branch office in Shenyang. Today, Mr. Schulz is leading the branch office in China as a project manager and principal architect. He has completed medium and large projects, among which are airports, stadiums, administration centers, hospital, office and residential designs. Each project designed in China includes ideas and expertise of ecological and sustainable design (ESD). With the issue of this book, he hopes not only to encourage designers to consider ESD in their designs, but also to inspire developers and investors of reconsidering ESD when taking actions and responsibilities for the future development. The book gives ideas of combining design and technical aspiration into the great architecture.

INNOVATION & BALANCE
RESIDENT ARCHITECTURE
创新与平衡·住宅建筑

With the strong appeal from people on "low-carbon life", the rapid development of green economy and environmental protection has been representing the general trend worldwide. In particular, people increasingly focus on construction energy consumption accounting for about one third of the total energy consumption globally. As a result, with the beautiful desire of protecting habitat for humanity together, german senior architect Daniel Schulz elected as editorial adviser, editorial board of the book collaborate with authority in the field of architectural design all over the world, planning and compiling this series of classic work together. The work aims to enlighten readers and make them reflect on Ecological Habitat, thus initiate solutions for harmonious commensalism on human beings, architecture and environment.

Taking green design concepts such as energy saving, low carbon, environmental protection, reuse and sustainability as the main theme, this book has selected nearly 100 recent outstanding green architectural works worldwide. A wide variety of green works and perspectives are included, ranging from public buildings to private residences, from high-end architectural complex to affordable social housing, from strategies of passive energy saving to techonologies of zero energy consumption. Along with specific planning drawings and professional conceptual introduction, readers are able to have a deep insight into how creatively the renowned architects express and interpret green concept. Meanwhile, professional readers can also aquire stronger competitive advange in this field and expand their business opportunities by studying these remarkable green design ideas in this book.

随着人们对"低碳生活"的强烈呼吁，环保、绿色经济的迅猛发展已经成为全球大势所趋。其中，约占世界总能耗1/3的建筑物能耗更日益成为人们关注的焦点。由此，秉承"携手保护人类家园"的美好愿景，本书编委与各国建筑设计界权威人士倾力合作，并推举德国资深建筑师Daniel Schulz担任编辑顾问，共同策划编写了这套经典著作。希望能够借助此书带给人们更多关于生态人居的启发和思考，从而提出更多人类、建筑与环境和谐共生的解决方案。

本书以节能、低碳、环保、再利用、可持续性等绿色设计为主题，为读者精选出近100例新近期最为突出的绿色建筑作品。从公共建筑到私人住宅，从高端建筑群到社会保障房，从被动式节能措施到零能耗领先技术，在书中都有所涵盖，同时配以详细的设计方案图和专业的理念解读，使读者可以充分领略到世界各地知名建筑师如何创造性地表现与诠释绿色理念。与此同时，专业读者通过对本书绿色创意的研究，将获得更大的行业竞争优势，利于拓展更多的商业合作机会。

CONTENTS

MIX-USE

Twelve | West Mixed-Use Building	8
8 House in Copenhagen	14
Bumps in Beijing	26
Le Monolithe in France	32
Sluňákov in Czech Republic	36

HOUSE

BayShore House in USA	44
Cher House in Argentina	52
City View Residence	60
Clayton Street Residence	68
Darlington House in Australia	72
Deepstone in UK	76
Elling House in Russia	82
Ellis Residence in Bainbridge Island	86
Fab Lab House in Spain	92
Hillside House in Mill Valley	100
House in Nikaia	108
Iseami House in Costa Rica	116
Knowles Residence in Seattle	124
Kona Residence in USA	130
Kyneton House in Australia	140

IN GREEN !
RESIDENT ARCHITECTURE

Architectures with Energy Saving, Low Carbon,
Environmental Materials, Technology and Reuse Ideas

Kuhlhaus 02 in USA	144
MC1 Residence in Costa Rica	154
Mosman Green in Australia	162
Hidden House Residence	172
Ormond Esplanade House	178
Panel House in USA	188
Rozelle Green in Australia	194
Setia Eco Villa in Malaysia	200
Sideris House in Greece	206
The House The Slope in Russia	212
Sorrento Beach House	220
Stony Point House in USA	224
Telescope House in Russia	232
The Conservatory House in Varna	236
The Houl in UK	244
The Rent House in Moscow	254
Villa BH in the Netherlands	262
Villa Nyberg in Central Sweden	270
White Shade in Australia	274

APARTMENT

Da Vinci Residential Tower	282
Sierra Bonita Affordable Housing	290
Soe Ker Tie House	296
Urban Lake Housing in Italy	302

Twelve | West Mixed-Use Building

Rising 22 stories above Portland, Oregon's evolving West End neighborhood, Twelve | West is a mixed-use building designed to meet two LEED Platinum Certifications and serves as a laboratory for sustainable design and workplace strategies. It features street-level retail space, 4 floors of office space for the architectural firm, 17 floors of apartments and 5 levels of below-grade parking.

The building has an eco-roof, rooftop garden and terrace space, complete fitness studio and a theatre. Four wind turbines sit prominently atop the building representing the first U.S. installation of a wind turbine array on an urban high-rise. Twelve | West serves as not only an anchor in a rapidly transforming urban neighborhood, but also as a demonstration project to inform future sustainable building design. The project is a catalyst for the next wave of redevelopment in this city's resurging downtown.

Twelve | West 位于波特兰市，与发展中的俄勒冈西区相邻，是一栋混合用途的大楼，其设计符合两项LEED铂金证书，还被作为可持续性设计和工作场所策略的实验室。沿街的一层是零售区，建筑公司占据了4层办公区，其余的17层是公寓，地下还有5层停车场。

大楼还配备生态屋顶、空中花园、露台、健身房和一家影院。楼顶的四台风力涡轮机十分醒目，它们还是首次出现在美国的城市楼顶，Twelve | West 不仅是这个日新月异的城市中的一个港湾，还是展示未来可持续建筑设计的一个重要工程。这一工程将对城市中心的新一轮改造起到促进作用。

Name of Project / 项目名称:
Twelve | West Mixed-Use Building
Location / 地点:
Portland, USA
Area / 占地面积:
51,096 m²
Completion Date / 竣工时间:
2009
Architecture / 建筑设计:
ZGF Architects LLP
Interior Design / 室内设计:
ZGF Architects LLP
Photography / 摄影:
Eckert & Eckert,
Timothy Hursley,
Nick Merrick © Hedrich Blessing,
Basil Childers,
Sherri Diteman
Client / 客户:
Gerding Edlen Development Company

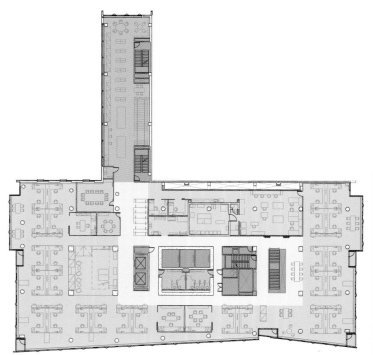

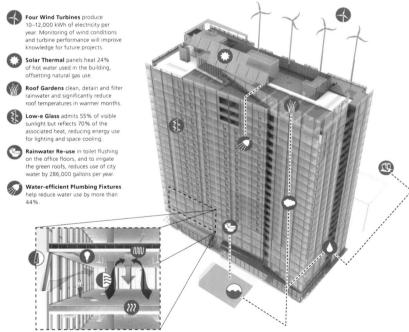

- **Four Wind Turbines** produce 10–12,000 kWh of electricity per year. Monitoring of wind conditions and turbine performance will improve knowledge for future projects.
- **Solar Thermal** panels heat 24% of hot water used in the building, offsetting natural gas use.
- **Roof Gardens** clean, detain and filter rainwater and significantly reduce roof temperatures in warmer months.
- **Low-e Glass** admits 55% of visible sunlight but reflects 70% of the associated heat, reducing energy use for lighting and space cooling.
- **Rainwater Re-use** in toilet flushing on the office floors, and to irrigate the green roofs, reduces use of city water by 286,000 gallons per year.
- **Water-efficient Plumbing Fixtures** help reduce water use by more than 44%.

- **Operable Windows** provide occupants fresh air, cooling, and a connection to the outdoors.
- **Daylight Sensors** switch off electric lights when there is ample daylight, reducing lighting energy by 60%.
- **Exposed Concrete** moderates indoor air temperatures. Mass is cooled with cool night air in the summer months and absorbs excess heat throughout the day.
- **Passive / Chilled Beams** provide energy-efficient cooling on the hottest days.
- **Under-Floor Air Distribution** efficiently delivers moderate-temperature air directly to occupants. Personal adjustable floor vents provide control over ventilation.
- **Water Storage Tank** temporarily stores up to 22,000 gallons of rainwater and condensation for re-use.
- **Efficient Central Cooling** plant in the nearby Brewery Blocks provides chilled water for space cooling.
- **Rain Water Harvesting** piping gathers 273,000 gallons of rainwater from the roofs.
- **Condensation** of 13,000 gallons of water from the air handler system will collect during summer months.

The design team made the bold decision to incorporate building-integrated wind turbines into the design of Twelve | West, a process that led not only to the mounting of four wind turbines on the tower's roof but, more importantly, to a body of research that will help advance the application of building-integrated wind power and provide a roadmap for other projects.

The turbines are predicted to generate roughly 10,000 to 12,000 kWh per year, enough to power the elevators. More importantly, the turbines will be thoroughly instrumented so that actual performance and wind flow patterns can be validated against predictions. Recognizing the ground-breaking rigor of investigation into this untested application, the Energy Trust of Oregon and the Oregon Department of Energy funded the entire system cost through energy efficiency grants and tax credits.

设计团队大胆地将建筑一体的风力涡轮机这一设计运用在Twelve | West项目中。这一设计的重要性在于不仅为建筑一体的风力电能应用提供研究，并且还可以为其他立项开创先例。

根据预计，涡轮机每年发出的电能为1万到1.2万千瓦时，足够为电梯供电。更重要的是，涡轮机完全以仪表进行测量，从而使其实际性能和风动情况都可针对预测进行验证。认识到这将是一项开创性的探索，俄勒冈能源信托基金和俄勒冈能源部通过节能补助和税收抵扣的方式为整个系统提供了资助。

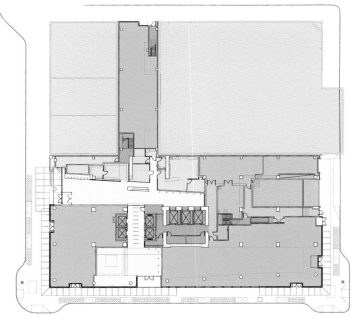

8 House in Copenhagen

8 House is not only offering residences to people in all of life's stages as well as office spaces to the city's business and trade – it also serves as a house that allows people to bike all the way from the ground floor to the top, moving alongside townhouses with gardens winding through an urban perimeter block.

The building's housing program offers three kinds of accommodation: apartments of varied sizes, penthouses and townhouses. The architects have designed a long, coherent house with immense differences in height, creating a strong inflow of light and a unique local community with small gardens and pathways that channel people's thoughts into mountains in Southern Europe and memories of a childhood home.

Instead of dividing the different functions of the building – for both habitation and trades – into separate blocks, the various functions have been spread out horizontally. The apartments are placed at the top while the commercial programme unfolds at the base of the building.

8 House不仅为各行各业的人们提供居住空间，还为市内商贸企业提供办公场所。同时，它也是一所房子，人们可以从底层骑车一直到房顶，或者在城市街区的花园洋房间穿行。

大楼可以提供三种居住模式：大小不等的公寓、顶层阁楼和独立洋房。建筑师们设计的这个房子狭长而辗转相连，同时在高度上相差悬殊。这种设计既保证了充足的自然光照，也使这个带有小花园和街道的当地社区风格更为独特。这一切可以将人们的思绪带到南欧连绵的山脉，或是童年家乡的回忆。

建筑中不同的功能区——居住区和商业区——并没有被分隔开来，而是依横向排列在一起。公寓被安排在顶部，商业区则占据此建筑的下层空间。

Name of Project / 项目名称:
8 House
Location / 地点:
Copenhagen, Denmark
Area / 占地面积:
61,000 m²
Completion Date / 竣工时间:
2010
Architecture / 建筑设计:
BIG
Landscape / 景观设计:
BIG
Photography / 摄影:
Dragor Luftfoto,
Jens Lindhe,
Ty Stange
Client / 客户:
St. Frederikslund Holding

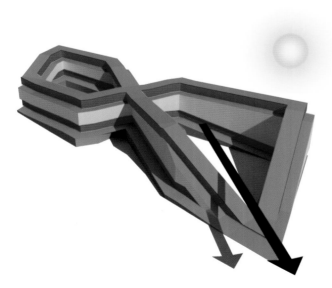

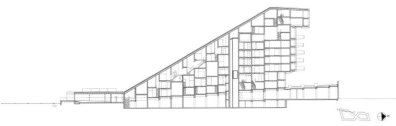

As a result, the different horizontal layers have achieved a quality of their own: the apartments benefit from the view, sunlight and fresh air, while the office leases merge with life on the street. This is basically emphasized by the shape of 8 House which is literally "hoisted up" in the northeast corner and "squeezed down" at the South West corner, allowing light and air to enter the court yard in the middle.

A continuous public path stretches from street level to the penthouses and allows people to bike all the way from the ground floor to the top, moving alongside townhouses with gardens, winding through an urban perimeter block. Two sloping green roofs totaling 1,700 square metres are strategically placed to reduce the urban heat island effect as well as providing the visual identity to the project and tying it back to the adjacent farmlands towards the south.

因此，不同的横向楼层都能从中获益：人们在公寓中可以享受户外景观、阳光和清新的空气；办公区也融入了街区的生活中。这些都得益于8 House独特的外形设计：东北角高耸，西南角压低，从而使更多光线和空气进入中心庭院。

一条公共街道直通顶层公寓，人们可以一直骑车上去，在花园洋房和城市街区中穿行。精心规划的两面坡状绿色屋顶总面积达1700平方米，它可以降低城市热岛效应，也是这座建筑的视觉识别所在，还巧妙地将建筑与南面相邻的农场相连。

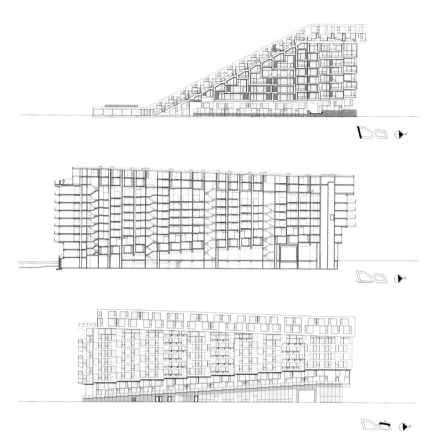

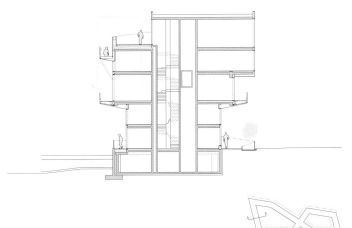
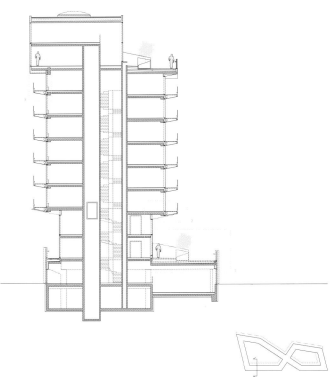

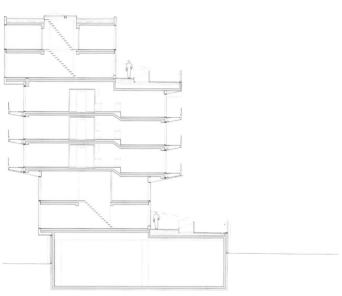
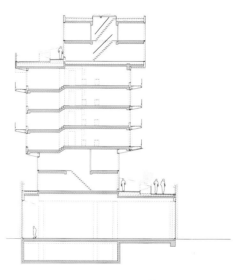

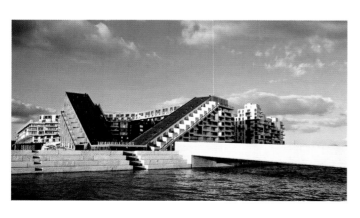

IN GREEN！RESIDENT ARCHITECTURE • MIX-USE

Bumps in Beijing

Based on Beijing's context, "Bumps in Beijing" was designed to have rectangular form and random courtyard, and the façade was placed random windows on rigid grids to maintain the same style.

The residence buildings are 80 metres in height. Every two floors are set as a unit. Every unit is staggered by 2 metres horizontally. The whole building of 80 metres is repetitions of these units. The setback areas are used for terraces. Interlaced black and white units highlight the concave-convex facade and show a clear lineament of the building. Windows, mirror windows and air conditioning units louver windows are embedded into exterior walls. All the windows are 1 square metre size. Randomly placed windows weaken the existence of all pillars and beams. Therefore, the appearance of the buildings look like stacked by lots of small black stone and white stone "boxes" together.

Tall buildings usually emphasize their verticality. "Bumps in Beijing" however emphasizes its horizontality. The 8-floor commercial building also incorporates some of the staggered terraces. Also, the beams extend 9 metres from the facade underscoring the project's dynamic business functions. "Bumps in Beijing" is not to merge into surrounding circumstances, but a "strong existence" as a landmark leading the development of the site. At the same time, the staggered black and white "boxes" and random windows units form energetic and dynamic complex images. It weakens the building volumes, and enlarges the building expressions.

Name of Project / 项目名称:
Bumps in Beijing
Location / 地点:
Beijing, China
Area / 占地面积:
103,218 m²
Completion Date / 竣工时间:
2008
Architecture / 建筑设计:
SAKO Architects,
Keiichiro Sako,
Yoko Fujii,
Jiye Zhang,
Hiroaki Sawamura
Photography / 摄影:
Beijing Ndc Studio, Inc.,
Misae Hiromatsu
Client / 客户:
Beijing Xinfengxinde Real-Estate,
Development Co., Ltd.,
Beijing Zihexin Plaza Co., Ltd.

基于北京当地的建筑环境，"北京冲击"被设计成长方体构造结合不规则庭院的形式，外墙坚固的网格上装有随机分布的窗户，以保持整体风格的和谐统一。

住宅楼高80米。每两层为一个单元。每个单元横向错开两米。整个80米高的建筑不断重复应用这些单元。楼宇外墙缩入的部分用做露台，黑白相间的单元色彩突出了外墙的凸凹感，也使建筑轮廓更为明晰。窗体、镜面窗、空调通风窗都被嵌入外墙，所有窗体面积都为一平方米，其分布上的随意性弱化了立柱和横梁的存在感，使建筑看起来像是由许多黑色和白色的小石"盒子"堆叠而成。

高层建筑通常着重表现纵向，但本案却着重强调其水平面。8层高的商用建筑也穿插着一些交错的露台。延伸至墙外9米的横梁更凸显了项目的动态商务功能。"北京冲击"并没有刻意去融入周边环境，而是有望成为一个地标性的"强大存在"，引领该区域的发展。交错的黑白"盒子"和不规则的窗户组合勾勒出充满活力、动感的一体画面。在弱化了建筑体感的同时，却强化了建筑的感染力。

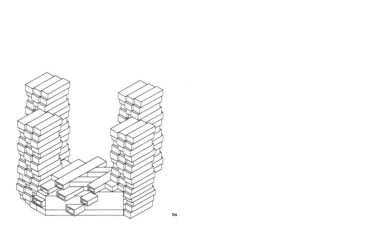

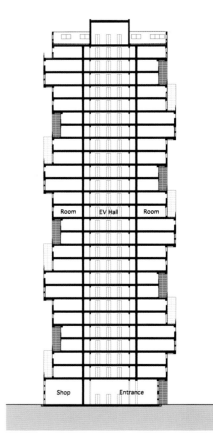

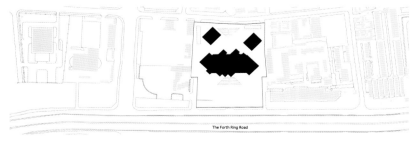

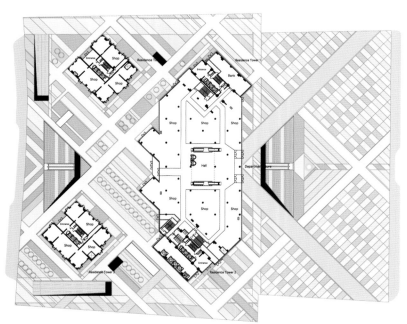

Beijing is built along its cardinal axis, and its most city districts reflect this kind of rectangular form. "Bumps in Beijing" is an integrated energy-saving project with four residences as well as a commercial building. Together, their density exceeds 350%. The traditional residence buildings in China are oriented south and north. With the increase in the density of the buildings, the traditional method causes buildings too close to each other and the rooms facing to the north can hardly get sunshine.

"Bumps in Beijing" is rotated by 45 degrees from the north-south axis for increased daylighting. The unique positioning of "Bumps in Beijing" project can save a considerable amount of energy. This design can provide optimum sunshine for each building and also can short cut the way to the central business areas from different directions. With staggered boxes and 45 degrees rotated from north-south axis, its varied building profile changes with location and also enriches the varying of the sunlight and shade .

Most of the households have balconies, thus easy to get communication with outside and get sunlight and wind. Most of façade material is stone, which can save energy compared with glass curtain wall. In addition, stone façade is easy to maintain very much. There is not fence of the project, so it can bring in the surrounding environment. "Bumps in Beijing" is a good rest plaza for the residents in the site and the surrounding.

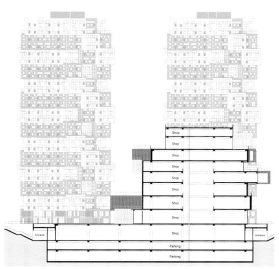

北京是以轴线为基础而建造的城市，多数城区都沿其方向呈长方形分布。由四幢住宅楼和一幢商用楼构成的"北京冲击"是一项节能工程，其容积率超过350%。中国传统的居民楼都是坐北朝南建设的正向建筑。随着建筑密度的增大，传统形式会导致楼间距减小，从而使朝北的房间难以享受到日照。

为增加日照量，"北京冲击"地块的南北轴线被旋转45度。独特的方位设计可以节约相当一部分能源。这种设计可以使每栋建筑得到最优化的光照，还可以使从不同方向到达中间商业区的路程最短化。交错的盒子加上45度旋转，使得建筑物轮廓随着方位变化而改变的同时，也丰富了光影上的视觉变化.

大部分户型都设有阳台，便于与外界交流，同时享受日光与微风。建筑外墙主要采用石材料，与玻璃幕墙相比，更利于节能。并且，石质墙体也易于维护。项目四周没有安设护栏，以便使外围环境与其自然相融。对于该区域和周围的居民来说，"北京冲击"是一处极佳的休闲广场。

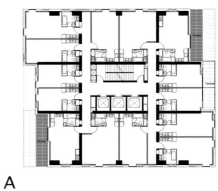

A

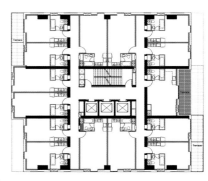

B

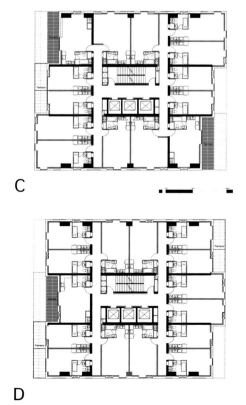

C

D

Le Monolithe in France

"Le Monolithe", an energy efficient mixed-use urban block located in the development area confluence at the southern tip of Lyon's Presqu'île, has reached completion. The structure with a total surface of 32,500 square metres combines social housing, rental property, residence for disabled people, offices and retail.

The block is composed of five sections, each one designed by a different architect, following the MVRDV masterplan: Pierre Gautier, Manuelle Gautrand, ECDM and Erik van Egeraat. Landscape architects "West 8" designed the public plaza. MVRDV designed the head section which advertises over the full façade the European integration by quoting the EU constitution. "Le Monolithe" has been realized by ING Real Estate Development and Atemi.

Interiors of the south facing building are protected from the sun by means of aluminium shutters as a reference to traditional local architecture. Apartments inside Le Monolithe offer a great diversity in order to attract different groups of inhabitants making the block a reflection of Lyon's population. Offices are divided into separate units of min. 500 square metres which are accessed by three vertical circulation cores, providing individual access. Each unit allows for a flexible fit out, depending on the tenants' needs and requirements. All spaces are naturally lit and ventilated.

Name of Project / 项目名称:
Le Monolithe
Location / 地点:
Lyon, France
Area / 占地面积:
32,500 m²
Completion Date / 竣工时间:
2010
Architecture / 建筑设计:
MVRDV
Landscape / 景观设计:
MVRDV
Photography / 摄影:
Philippe Ruault
Client / 客户:
ING Real Estate Developers

位于法国里昂Presqu'île南端的开发区交汇处，一座高效节能的城市综合体项目"Le Monolithe"已经建成。这个表面积为32 500平方米的建筑群包括社会保障房、出租房、残疾人住房、办公楼和零售店。

建筑群由五部分组成，按照MVRDV团队的总体规划，每部分都由不同的建筑师设计，他们分别是：Pierre Gautier、Manuelle Gautrand、ECDM和Erik van Egeraat。公共广场由景观建筑事物所"West 8"主持设计。MVRDV团队负责设计的主题部分占据整面外墙，以推广由欧盟宪法引述的欧洲一体化。"Le Monolithe"项目由ING地产开发公司与Atemi公司合作建造而成。

铝质百叶窗的安装使得该南向建筑的室内部分能够免受阳光直射，这种方式借鉴于当地传统建筑。"Le Monolithe"内部的公寓风格各异，吸引了不同居民群体来此居住，从而使这里成为里昂人居状态的一种写照。办公区被分割成若干独立单元，最小面积为500平方米，人们可以通过三个主要垂直通道进入。每个单元可以根据住户的需要和要求进行调整。整个建筑都采用自然光和自然通风。

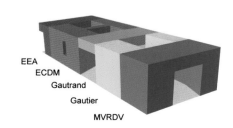

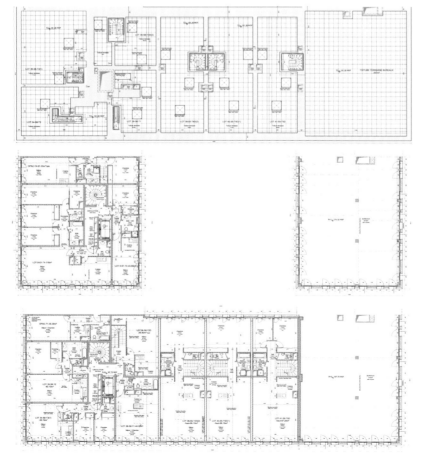

"Le Monolithe" is one of the projects within the greater scheme for Lyon Confluence which has been developed as part of Grand Lyon's European Concerto-Renaissance programme, a project supported by the European Commission. The building not only complies with High Environmental Quality (HQE) criteria, such as reinforced insulation, careful selection of materials and rainwater management; further, 80% of the total energy consumed is provided by renewable energy sources. The combination of efficient spatial composition, passive energy (sunscreens, high thermal inertia), thermal and acoustic comfort and an energy strategy that includes heat storage, PV-cells, low-e double glazing, compactness to minimise heat loss, natural ventilation and an environmentally responsive façade system make "Le Monolithe" a highly efficient low energy construction, e.g. heating accounts for <40 kWh/m²/year and hot water <5 kWh/m²/year.

"Le Monolithe" 是"汇聚里昂"这个更大主题下的项目之一，也是"大里昂欧洲和谐复兴"计划的一部分。该项目由欧盟委员会资助，建筑不但符合高级环境质量标准(HQE)，如加强绝缘、谨慎选材以及雨水处理，除此之外，所耗能源的80%都由再生能源担负。高效的空间构成、被动能源(遮光剂及高热惯性)、热度和声环境舒适，和能源战略被综合运用于此，其中包括：储热、光伏电池、低辐射双层玻璃窗、密封以减少热量流失和自然通风，以及根据环境调节的外墙系统。这一切使"Le Monolithe"成为高效节能建筑的典范。其热量消耗不超过40千瓦时/平方米.年，热水消耗少于5千瓦时/平方米.年。

Sluňákov in Czech Republic

The building has been designed as a curved inhabited land wave that fluently blends into the surrounding terrain. The architectural design utilizes the southern orientation with a southern glass facade with movable sun blinds. Two recessed entrances are situated at the north side.

The eastern side of the building symbolically ascends from the ground to enhance the display of the southwestern sunshine. The earth-sheltered northern side of the building fluently adjoins the building's earth-covered roof, which gradually increases in height from the west to the east. The path that leads the visitor from the main entrance over the "ridge" of the building goes on to the "top" lookout point that provides a view of the entire biocenter.

The backbone of the building is the hallway that runs the entire length of the building. On the sunny south side from the hallway is residential area, partially one floor (lecture hall, dining room, class rooms, offices), partially two floors (accommodation, governors flat) and on the shaded north side there are appliances.

Name of Project / 项目名称:
Sluňákov – Centre for Ecological Activities
Location / 地点:
Horka nad Moravou, Czech Republic
Area / 占地面积:
1,586 m²
Completion Date / 竣工时间:
2006
Architecture / 建筑设计:
Projektil Architekti,
Roman Brychta,
Adam Halíř,
Petr Lešek,
Ondřej Hofmeister
Photography / 摄影:
Andrea Thiel Lhotáková
Client / 客户:
City of Olomouc, National Found of the Czech Republic for Environment

该建筑被设计成弧形居住区，与周围地形完美地结合在一起。建筑南面采用了配有可移动遮光板的玻璃外墙。两个缩进式入口都位于北面，建筑东侧象征性地高于地面，以加强西南方阳光的照射。被地面覆盖的建筑物北侧与同样被覆盖的屋顶相连，高度由西向东递增。一条小路引领访客从主入口到建筑物的"脊梁"、再到达"顶端"的瞭望点，在那里可以一览生物中心的全貌。

建筑的骨干部分是贯通整个建筑的门厅。门厅南侧阳光灿烂的一面是居住区，其他部分有的位于一层（演讲厅、食堂、教室、办公室），有的占据二层（住宿、政府公寓），在北侧背光的一面则是各种配套设施.

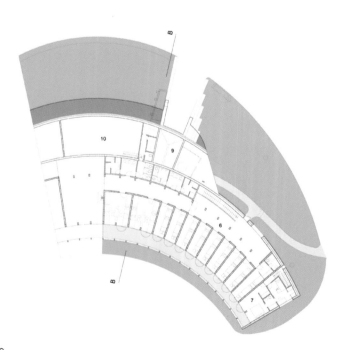
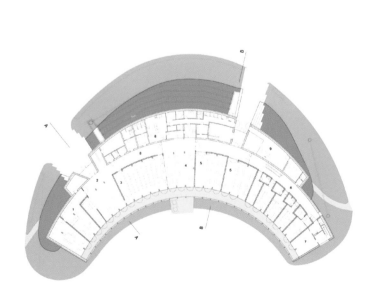

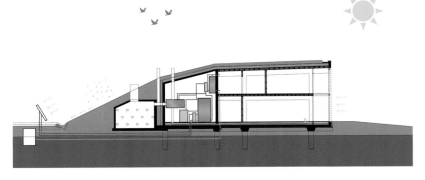

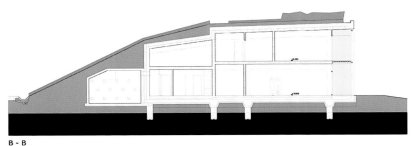

B - B

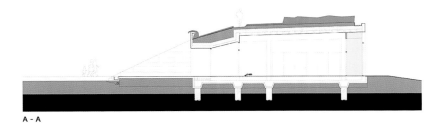

A - A

All materials used in the project have been chosen due to their environmental friendliness. The entire concept of the interior and exterior of the building takes into account the use of natural colors and surface structures of the individual building materials. The facades are covered by wood, glass, concrete and stone (stacked). The interior is completed using mainly wood, glass and unfired brick walls and plaster.

The building-energy concept was designed with respect to the basic principles of sustainable development. The building is designed for full yearlong operation. The heat demand is covered using a combination of renewable energy sources – biomass and solar energy. Ventilation and warm air heating are ensured by fresh air and by warm air circulation ventilation with heat recovery from outgoing air.

Earth heat exchanger serves mainly to bring cooler air inside in the summer months. Two automatic wood pellet furnaces provide the main source of heat for heating and are a supplementary source for heating the hot water. A modern solar system that decreases the need for secondary energy for hot water heating has been proposed.

建筑中所有用到的材料都是基于环保原则而选。建筑内部和外部的整体理念都采用天然的色彩和独立建筑材料的表面结构。外墙由木材、玻璃、混凝土和石材(层叠的)进行装饰。内部则主要使用木材、玻璃、未经烧制的砖墙和石膏。

建筑能耗理念遵循可持续发展的基本原则，按照建筑物整年运作进行预算。对热量的需求主要通过生物和太阳能两种可再生能源综合供给。通风与空气加热则通过新鲜空气以及回收流出空气的热循环来实现。

地热交换器的主要功能是在夏季时将较凉爽的空气引入室内。两个自动木屑颗粒炉是热量的主要提供者，并负责补充采暖和水加热的能耗。一个能够降低对于二次能源需求的现代化太阳能系统也在计划之中。

044-279

HOUSE

RESIDENT VOLUME

PART 2

BayShore House in USA

The Balinese style of architecture exemplifies a strong link between the indoors and outdoors and is achieved in this house through the use of a series of high french doors that open directly to heavily shaded porches on the second floor and to the pool/patio on the first floor.

The swimming pool and patio area are the main focus of this house. In keeping with the Balinese design concept of the home the interface between house and the outdoors is prominent. The architects have tried to make this area a true focal point for viewing as well as activity. The geometric modern design of the pool and pattern scored concrete deck is exemplified again three dimensionally with the walkway and sidewall leading to the guest suite. The green glass panels are lighted at night to not only highlight the avocado tree, but in so doing, light the walkway at the same time.

The architects experimented with the low stucco walls at the pool area by mixing a small shell into the plaster mix to render a textured wall that matches the shell impregnated low walls at the front of the house, but is less prone to mildew buildup. The two waterfalls, both lighted at night, drop into shallow swimming areas, and are focal points from the master bedroom and the living and family rooms.

巴厘岛建筑风格常表现为室内、外空间的紧密关联。通过运用一系列高大的法式门，直接开向二楼的遮阳门廊或者一楼的游泳池／露台，使得巴厘岛建筑特色在本案设计中被成功实现。

游泳池和露台是该住所的焦点区域，要将打造巴厘岛建筑风格的理念贯穿到底，房子和户外空间的接合处设计十分重要。建筑师努力将这一区域设计成视线范围和活动空间的聚焦点。富有现代感的几何形游泳池和精雕细刻的混凝土露台在通往"客房"的走廊和侧壁的衬托中，其立体感得到再次烘托。绿色的玻璃镶板可以在夜间发光，这样不仅突出了鳄梨树的美丽，同时也照亮了走廊。

建筑师尝试在泳池区域低矮的墙壁上将石膏与小贝壳混合，渲染出纹理明晰的质地，与房前的贝壳矮墙互为映衬，同时也避免了发霉现象的产生。夜幕降临时，灯光倾洒在两个小瀑布上，水流缓缓地倾泻到水池里，从室内主卧房和起居室角度望去，这里更是一处设计亮点。

Name of Project / 项目名称:
BayShore House
Location / 地点:
Florida, USA
Area / 占地面积:
334 m^2
Completion Date / 竣工时间:
2008
Architecture / 建筑设计:
Terry G Green, Architect
Landscape / 景观设计:
Lynne Pitts
Interior Design / 室内设计:
Terry G Green, Architect
Photography / 摄影:
Matt McCartney
Client / 客户:
Speculation

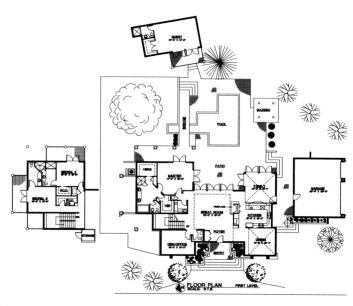

The original inspiration came from the Balinese concept of detached "Bale" pavilions – an architecture form representative of separated living zones (cooking, eating, praying, sleeping, etc.) as well as necessary separation for environmental reasons, e.g. cross ventilation, maximizing natural light, and separation of incompatible uses, (cooking and bathing, or sleeping and living rooms). In keeping with this concept the garage is detached (noxious fumes) from the house as well as a separate guest suite for privacy.

The house is certified by the Florida Green Building Coalition mandating that only sustainable building practices, materials and healthy finishes have been used. The most efficient, energy saving appliances and air-conditioning systems, and solar hot water heating, have also yielded a certification by Energy Star for Homes. Other important features include a state-of-the-art contemporary kitchen with synthetic stone countertops and beautiful sustainable bamboo cabinetry and stair treads and hand rails, sloped cathedral ceilings in living, dining and family rooms, and "art" niches, throughout the house, to display one's personal treasures.

本案设计的最初灵感来源于巴厘岛建筑中"Bale"式分离概念，那是一种强调各功能区（厨房、用餐、祷告和卧室等）相互分离的建筑形式，是受环境因素影响而产生的一种设计理念。例如，良好的对流通风、充分地利用自然光线以及将毫不

相干的使用区域分离开来（烹饪和沐浴，或者卧房和客厅）。为了贯彻这一理念，车库（产生有害气体）和主建筑被分离开来，考虑到私密性，客房也被单独安设在一处。

这所房子已经过佛罗里达州绿色建筑联盟认证鉴定，将可持续性建筑措施、材料和健康环保的装饰应用于本案设计。最有效的节能设施、空调系统和太阳能热水供暖也同样通过了能源之星家庭体系的认证。其他的可持续设计还包括一个国内最先进的、带有人造石工作台面的现代化厨房，由可持续性竹子制成的精美橱柜、楼梯踏板、扶手和教堂式的斜顶天花板，房屋里展示个人财富的艺术壁龛也同样具有可持续性。

Cher House in Argentina

A couple with two teenage children had commissioned María Victoria Besonías and Luciano Kruk of BAK Architects to build them a home for the summer. The household have approximately 113 square metres to be tightly integrated into the landscape and take advantage of spectacular views to the nearby forest, and the chosen location is Mar Azul, Villa Gesell party, province of Buenos Aires, Argentina. This beautiful house is suitable for the family who love challenge and adventure.

The theme of the house is back to nature. The interior and exterior of the beautiful house is made from nature material. The use of concrete extends to the furniture except the main bed, chairs and sofas, thus limiting the use of materials and minimizing the impact of the house in the area. The program has two bedrooms, two bathrooms, one en suite and a common place (with kitchen) as generous as possible.

The property is developed in a concrete and glass prism that highlights the different levels at which this resolved. The interior walls of hollow bricks are plastered and painted with white latex, and the floor cloths is divided by smoothing cement plate of aluminum.

一对夫妇及他们的两个十几岁的孩子委任BAK Architects团队中两位建筑师María Victoria Besonías和Luciano Kruk设计一处房屋，供家庭避暑之用。主人拥有将近113平方米的土地面积，并可充分利用周围森林景观的优势。房屋选址在阿根廷首府布宜诺斯艾利斯省的Mar Azul。这座美丽的林间房屋十分适合喜爱冒险与挑战的家庭居住。

本案设计的主题即回归自然，房子室内、外装饰均选用天然材料。家具除床、椅子、沙发外均以混凝土为主要材料，节省材料的同时也降低了对周围环境的影响。房内设有两间卧室、两间浴室、一间套房及带有厨房的公共区域。

混凝土及玻璃棱镜的运用凸显出不同的层次。内墙选材空心砖，涂以灰泥，外层粉刷白色乳胶漆。地面则以光滑的铝制水泥板进行铺制。

Name of Project / 项目名称:
Cher House
Location / 地点:
Buenos Aires, Argentina
Area / 占地面积:
113 m²
Completion Date / 竣工时间:
2009
Architecture / 建筑设计:
BAK Architects,
María Victoria Besonias,
Luciano Kruk
Photography / 摄影:
Gustavo Sosa Pinilla
Client / 客户:
Javier Daulte

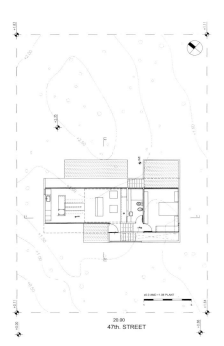

Recognizing and seizing the particular maritime-forest microclimate together with the low budget availability and with the requirement of a minimum posterior maintenance is determinant in the decision of the esthetic-constructive system of the house. The need to capture the light that passes through the trees leads architects to conceive the house as semicovered and resolve it with big windows that provide long hours of natural light and full integration with the scenery. While from the outside, the windows reflect the privileged landscape creating a house with less presence.

The conditions previously named and the need of accelerating the constructive stages, make architects choose the exposed concrete as the main material. They focus on low maintenance and use materials with a low budget. Furthermore, the shadow of the forest allows them to use this material since it provides the sufficient thermal insulation. The thermal conditioning of the house in winter does not matter much because it is supposed to be used in summer; however, the architects foresee a heating system. The water resistant insulation is resolved by using a very compact concrete and studying the shape of the outer-skin in order to reach a faster water drainage. On the other hand, the colour and texture of the exposed concrete have a strong and mimetic presence, so the work is in harmony with the landscape. The house is designed with passive sustainability criteria. It is adapted to

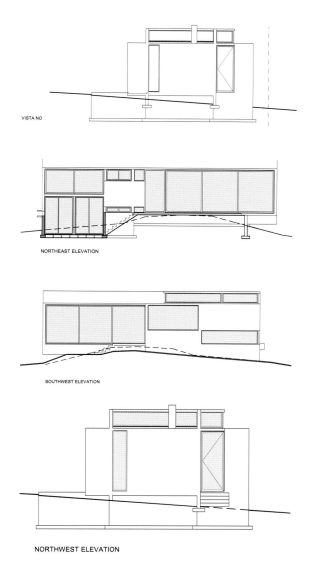

VISTA NO

NORTHEAST ELEVATION

SOUTHWEST ELEVATION

NORTHWEST ELEVATION

the environment where it is placed since it is surrounded by trees that protect the house, and the ideal of low impact in the environment comes true.

建筑师充分考虑到本案会受到海洋森林微气候的影响，打造出低预算的家居建筑，将维修费用降到最低，这对本案的美学构架系统起到决定性作用。建筑师为本案设计了巨大的窗户，便于林间的自然光线透过窗户射入室内，同时也与自然美景完美融合。从外观看，巨大的玻璃窗上映衬着周围景观，使整个建筑若隐若现。

客观条件及加快建筑过程的需要令建筑师主要选用混凝土作为房屋的主要材料。低维护费用和低成本的材料成为首选。而且森林遮阴具有良好的隔热效果，使混凝土的应用成为可能。该房屋主要用于夏天居住，因此冬天的温度问题并不重要。即便如此，建筑师还是为其设计了供暖系统。紧凑的混凝土结构保证较好的防水效果，外部构造利于快速排水。此外，房屋外观采用的混凝土在色彩、质地上的强烈视觉效果与森林风景自然融合。本案的设计符合被动式可持续发展准则，丛林掩映下的房屋与自然和谐共存，使对环境产生较低影响的愿景成为现实。

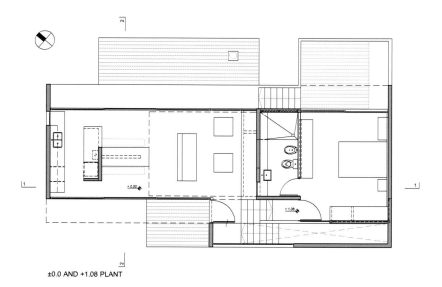

±0.0 AND +1.08 PLANT

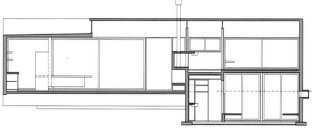

SECTION 1

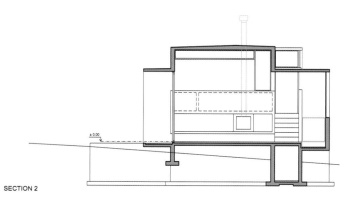

SECTION 2

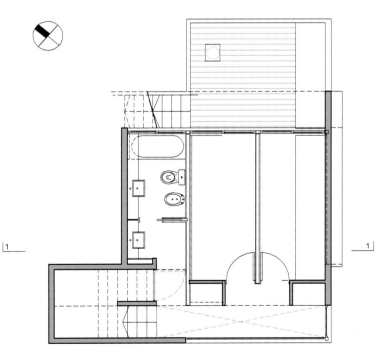

-1.47 PLANT

City View Residence

Located in Austin, City View Residence is a project which combine both modern design and sustainable elements, designed by Dick Clark Architecture, a company with a massive design portfolio from Texas, USA.

Inspired by the clients' desire to live in a house that reflects their values and lifestyle, the house takes an architecturally minimal approach with simple massing, an understated material palette, and large expanses of glass. Configured for entertainment, the glass box living area, along with large sliding doors from the dining and pool room invite guests outside onto the pool deck to take in the spectacular view of the Austin skyline.

The plan is very open, there are visual connections between the living/dining/kitchen. A see through oak screen separates the stair from the kitchen. To separate the public entry hall from the private bedroom wing, architects did a floating bookcase/art display millwork piece, and the design team used an ecosmart burner as well, which allowed architects to create a fireplace mass that doesn't touch the ceiing which adds non-partition quality of the house.

City View住宅位于奥斯汀，由来自德州的多产建筑公司Dick Clark Architecture担纲设计，住宅既采用了现代的设计方式，同时又融合了大量的可持续元素。

由于客户希望拥有一个可以反映自身价值观和生活方式的住宅，建筑师选择极简的建筑风格和低调的材料组合，同时运用许多大块玻璃元素。出于要接待客人的考虑，客厅采用了透明的设计，仿佛一个巨大的玻璃盒子。餐厅和台球间装有大的侧滑门，客人们可以在泳池边欣赏奥斯汀天际壮观的景色。

住宅采用了比较开放的设计形式，在视觉上，生活区、用餐区和厨房三区紧紧联系在一起。一面透明的橡木屏幕将楼梯和厨房分开。为了将卧室从公共入口厅分离出来，建筑师设计了一个悬浮木质结构，用来摆放书籍和陈列艺术品。同时，建筑师特别选用了EcoSmart节能燃烧器，创建了不触及天花板的壁炉，从而使空间不必有太多分区。

Name of Project / 项目名称:
City View Residence
Location / 地点:
Austin, USA
Area / 占地面积:
464.5 m²
Completion Date / 竣工时间:
2009
Architecture / 建筑设计:
Dick Clark Architecture
Interior Design / 室内设计:
Dick Clark Architecture, ABODE (Fern Santini)
Photography / 摄影:
Alex Stross,
Paul Bardagjy

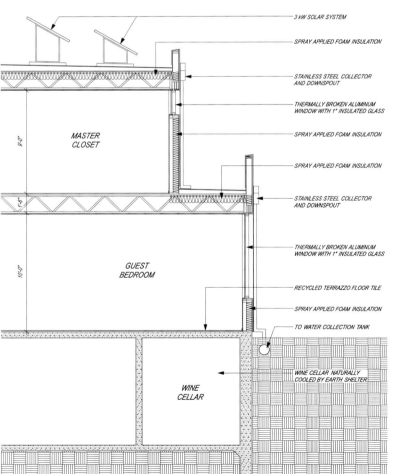

The architects tried their best to have as little impact on the environment as well. The house has a 3 KW solar panel system, instant hot water heaters, bio foam insulation, rain water collection that irrigates a good portion of the lot, very efficient mechanical systems, and low e-glass. The insulated glass with the deep overhangs minimized exposure to the sun and spray foam was used to keep the heat out. Eco-smart fireplaces and tankless water heaters reduce environmental impact while solar panels and rainwater collection utilize the resources the environment has to offer. Materials, such as recycled glass Terrazzo tile, cypress wood and concrete, were all sourced locally. All appliances are Energy Star Certified and the wine cellar in the partially subgrade basement requires no mechanical cooling.

建筑师竭尽全力把该建筑对周围环境的影响降到最低。房屋内安装有功率为3千瓦的太阳能板发电系统和即时热水器，解决了居家用电、用水问题，还应用了生物发泡保温技术。雨水采集系统可以将大部分采集而来的雨水用于植物灌溉。房屋内还安装了许多非常高效节能的机械系统和低辐射玻璃，有效地保护了室内和周围环境。隔热性高的中空玻璃的使用，以及屋顶的许多突出结构，大大地缩小了日照面积。同时，喷雾泡沫的使用也有效地阻挡了热量侵入。房内所用壁炉和热水器为生态智能型产品，减少了对环境的干扰。太阳能板和雨水采集系统的应用则提高了资源利用率。该建筑所用的许多材料，例如，可回收的水磨石瓦片、柏木和混凝土等，均在当地取材。室内所有家用电器均通过了能源之星认证。部分位于地下的酒窖更不需要任何机械制冷。

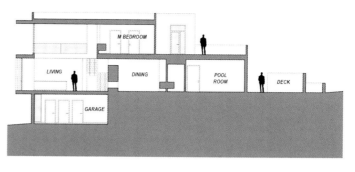
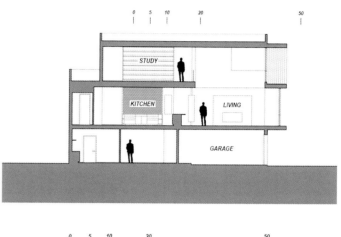

Clayton Street Residence

When purchased by the current owner, this 19th-century Victorian located in San Francisco's Haight Ashbury neighborhood was an eclectic mix of Victorian details and quirky California Hippie interventions. The original building had been extensively remodeled in the 1960's reflecting aesthetics and materials of that period, and a unique Victorian Hippie Vernacular had emerged. It was important to the owner that the new design maintained this unique patina while incorporating more natural light and enhancing the flow of the space to suit a more contemporary way of living.

Upon entering the home, the owner find himself in a light-suffused living room flushing with classic mid-century modern furniture, a reclaimed wood swing built by Law Nick Damner, and a beautiful staircase crafted from the building's douglas fir attic floorboards. These boards were salvaged, planed and sanded, and then glued together piece-by-piece to form a butcher-block stairway. Reclaimed wood from the attic was also used for the building's window trim and the flooring of a bridge that joins the residence's lofted spaces. A double-story space forms the heart of the residence and connects the two levels housing the public living space on the entry floor and more private sleeping quarters on the upper level. Dormers were added to the roof to accommodate more functional space and offer sweeping views of San Francisco landmarks.

Name of Project / 项目名称:
Clayton Street Residence
Location / 地点:
San Francisco, USA
Area / 占地面积:
260 m²
Completion Date / 竣工时间:
2008
Architecture / 建筑设计:
Mork-Ulnes Design
Interior Design / 室内设计:
Mork-Ulnes Design
Photography / 摄影:
Bruce Damonte

此业主购买的这幢建筑坐落于旧金山海德艾斯布利地区，始建于19世纪维多利亚时期。设计中融合了维多利亚风格的细节元素，兼有离奇的加利福尼亚嬉皮士风格。最初的房子经建筑师改造后，反映了20世纪60年代美学特色，材料运用十分丰富，一个独特的维多利亚嬉皮士风格住宅由此诞生。对房主来说，改造后的设计保留原有建筑中独特的经典之处这一点十分重要，同时要使房间能够吸取更多的自然光线，加强空间流动性，以符合现代生活的需要。

进入室内，首先映入眼帘的便是自然光线充足的客厅，其内摆放着别致的中世纪家具。由Law Nick Damner设计的秋千选用可再生木材料，漂亮的楼梯则由从建筑中回收的道格拉斯菲尔阁楼地板制成。这些木板经过回收、刨平、打磨等工序，再黏合到一起后被制成现在的楼梯。回收自阁楼的可再生木材也被制成窗户装饰，或用做连接共享空间的地板材料。双层的共享空间成为建筑中心并将楼下的公共生活区及楼上的私人休息区连接起来。屋顶天窗的设计不仅扩大了功能空间，同时也将圣弗朗西斯科的风景尽收眼底。

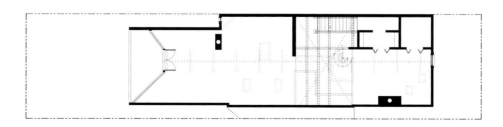

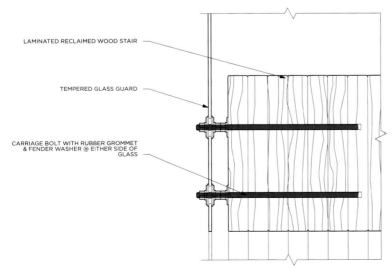

The owner, a designer and partner in the San Francisco Pre-Fab firm Modern Cabana, collaborated with his firm's team members in the construction and detailing. This lead to many of the unique design features, such as the recycled wood elements and a swing suspended from the steel bridge. A solid, butcher-block stairway and bridge was built using salvaged floorboards from the existing attic floor. Recycled and salvaged lumber and radiant heat floor contribute to the ethos of this former hippie flat.

The lofted areas contain the residence's living quarters. The space is punctuated by frequent skylights that bathe the airy modern interior in daylight, while the occasional exposed brick wall lets the building's past shine through. In carrying out the renovation Mork-Ulnes Design used recycled denim insulation within the walls, no-VOC finishes, and sourced the work of local cabinet makers. As a whole the home feels comfortably lived-in, vibrant, and full of history – it's a project that celebrates the past while looking towards the future.

业主作为旧金山Modern Cabana公司的设计师及合伙人，与公司其他成员在建筑和设计细节方面共同合作，打造出设计独特的建筑之作。如：可再生的木质元素、从钢梁上垂悬而下的秋千、由从建筑原有阁楼中回收的板材精制而成的坚固楼梯和共享桥梁、回收再利用的木材，以及地热设施，无一不将家居设计的嬉皮士风格演绎得淋漓尽致。

共享区域包括起居室。采光天窗的设计令此处空间沐浴在明亮的自然光线之中。偶尔外露的砖墙流露出建筑的几许历史沧桑感。设计团队在房屋改造中，将利于保温的可再生棉布置入墙体，并采用无VOC（挥发性有机化合物）饰面，同时由本地工作人员操作。总体来说，改造后的建筑居住舒适、富有活力，充满历史元素，堪称为承前启后的家居建筑之作。

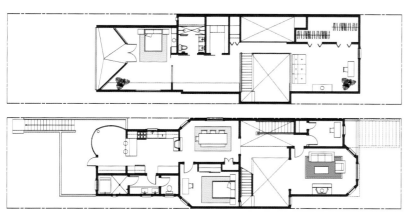

Darlington House in Australia

Located near Redfern in Sydney, the Darlington House was an opportunity for the architect to owner-build and experiment with a number of ideas that otherwise would not have been possible. Pivotal to the design was the concept that nature, even in an inner city environment, can permeate the everyday lives and that the design of the house should facilitate a connection with the natural elements.

The small site measuring 50 square metres was the greatest constraint and yet spurred the fundamental planning rethink to incorporate an outdoor bathroom. To achieve as spacious a living area as possible the original kitchen and living areas were swapped, making the kitchen the heart of the house, while the living area was positioned such that it borrowed a sense of space from the central courtyard. The bathhouse on the other side of the courtyard was connected to the courtyard by sliding screens that allow one to bathe outdoors when open. A translucent screen over the bathroom provides an organic architectural motif and privacy to living areas from the densely packed neighbourhood. The total floor space was increased to 55 square metres by converting the attic space into a loft room. The creative structural advice was provided by Bill Anderson.

Darlington House位于悉尼市雷德芬附近。通过它的改建，建筑师和业主得到了一个交流心得的机会，使许多不可能实现的构想成为了可能。设计的关键理念在于，即使在市中心，也要把自然风景融入到人们的日常生活中。因此，本案即需要与自然元素紧密相连。

虽然房屋建筑面积只有50平方米，但这一局限性却为设计师带来了灵感。他们在原有规划的基础上进行创新，设计出一个户外浴室。为了让居住空间保持宽敞，设计师还调换了厨房和起居室的位置，并把厨房设计成住宅的中心，起居室则充分利用了中间庭院内的很多空间。庭院旁边浴室与庭院之间隔有一道滑动屏风，这样人们就可以在户外享受淋浴。浴室顶部采用的半透明隔板，使其与周围分布密集的房屋相隔开，以有机建筑的形式打造出私密空间。顶楼被用做阁楼之后，总建筑面积被增加到55平方米。这些结构规划方面的创造性建议均由Bill Anderson提出。

Name of Project / 项目名称:
Darlington House
Location / 地点:
NSW, Australia
Area / 占地面积:
55 m²
Completion Date / 竣工时间:
2007
Architecture / 建筑设计:
Anderson Architecture
Interior Design / 室内设计:
Anderson Architecture
Photography / 摄影:
Steve Brown Photography
Client / 客户:
Simon & Kim Anderson

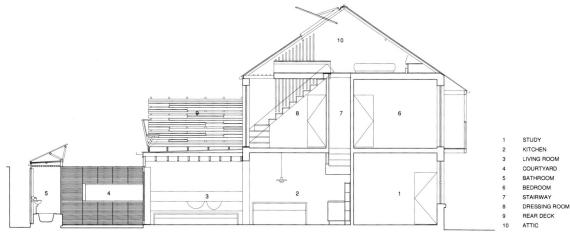

1 STUDY
2 KITCHEN
3 LIVING ROOM
4 COURTYARD
5 BATHROOM
6 BEDROOM
7 STAIRWAY
8 DRESSING ROOM
9 REAR DECK
10 ATTIC

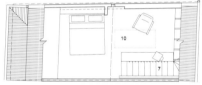

ATTIC

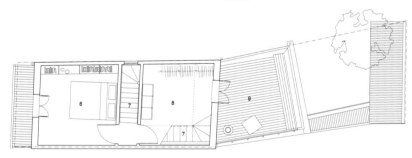

FIRST FLOOR PLAN

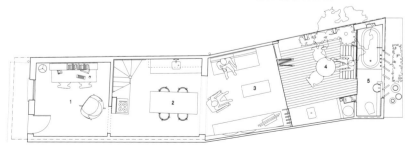

GROUND FLOOR PLAN

Close to 120 years old, the house has layers hoped may still be read and sustained, the ground floor stair remains in its original state, a sandstone threshold found between kitchen and living areas replaced, the uneven party wall revealed in the living room emphasised with hand-carved concrete seat back, and the raw rafters in the attic space exposed.

To optimise the house's sustainability recycled 40-mm thick Blackbutt timber, a sustainable yield hardwood, was used for the built in furniture such as the kitchen bench tops, living room bench and shelving. All internal finishes are zero volatile organic compound (VOC) paints or oil finishes improving indoor air quality. Other environmentally-friendly initiatives include a grey water reuse system, 3A rated water fixtures and efficient low wattage globes used in light fittings. Naturally easy to heat and cool due to its inherent size and redesign, this inner city worker's terrace is a valid ecological sustainable option.

该住所已有近120年的历史，为保持这里的历史底蕴和可持续性，一楼的楼梯仍保持原状，厨房和起居室之间的砂岩门槛则被替换掉。起居室内设有参差不齐的隔墙，衬托于此的是手工雕刻的混凝土椅背，和阁楼空间中暴露出的未加修饰的椽子。

为了优化房屋的可持续性，40毫米厚的莱克巴特再生木材和可持续使用的硬木被用在厨房台面、客厅长椅和搁架等室内家具的制作中。内部饰面都采用零VOC（挥发性有机化合物）涂料或者油画颜料，以此来提升室内空气质量。其他环境友好型设施还包括一个废水回收系统、3A级别的水装置和高效节能灯具。之前的加热、制冷设施在原有尺寸上也被重新设计。上述众多的环保设计措施的应用使这座位于市中心的工人露台成为一个不折不扣的生态环保型可持续项目。

Deepstone in UK

The house is located on a spectacular site overlooking the Solway Firth in Southwest Scotland. The site is a steeply sloping, former quarry in a National Scenic Area which slopes in two directions from the quarry base which forms the only level ground access.

The house is conceived as a stone plinth housing the bedrooms with a garage and entrance under at the level of the quarry base. The masonry base is finished in stone from re-cycled quarry waste. The principal living accommodation is expressed as a lightweight glazed "pavilion" sitting on the solid plinth. It is set back to form an external terrace facing the sea and to reduce the apparent mass of the house.

The glazed pavilion is constructed with a steel frame and highly insulated timber infill panels clad in cedar and triple glazed windows. The roof, although thick internally to provide very high levels of insulation, is cantilevered on all sides with projecting expressed douglas fir rafters to give a thin, elegant leading edge. The roof is finished in standing seam pre-fabricated grey zinc. The roof pitch follows the slope of the site to reduce the mass of the house. This also keeps the solar gain to manageable levels (the site faces due east) and the resultant section provides an outward sea view and an upward view of the landscape behind the house.

本案位于苏格兰西南部一处景象壮观的地方，从这里可以俯瞰苏格兰西北部的索尔韦湾。这里曾是国家风景区中的采石场，地势极为陡峭，从采石场的基地处开始向两个方向倾斜，建筑所在位置成为该区唯一一块平地。

住宅一楼位于采石场底层，设有车库和房屋入口。一楼由坚固的石材建成，仿佛一个庞大的石质基座。该底座所用石材源于原采石场废料。二楼是主要的生活空间，看起来则像是一个坐立在坚固石基上的轻巧玻璃亭。二楼空间略微向里，以建构面对大海的户外阳台，同时还能有效地保护住宅隐私。

二楼玻璃亭空间由一个钢制框架、高度绝缘并覆有雪松的填充板材以及三层玻璃窗构成。屋顶隔热性极高，尽管从内部看上去十分厚实，却采用悬臂式设计，以凸显花旗松椽，从而形成轻薄、优雅的屋檐。屋顶由预制的立式接缝灰色锌饰面。屋顶坡度顺和地形，有效地保护了室内隐私。这种设计还可将太阳能的获取量保持在可控范围内，同时便于业主欣赏海景和周围迷人的自然景观。

Name of Project / 项目名称:
Deepstone
Location / 地点:
Scotland, UK
Completion Date / 竣工时间:
2009
Architecture / 建筑设计:
Simon Winstanley Architects, principal – Simon Winstanley
Project Architect / 项目建筑师:
John Murray
Landscape / 景观设计:
Paterson Landscape
Interior Design / 室内设计:
Simon Winstanley Architects

first floor plan

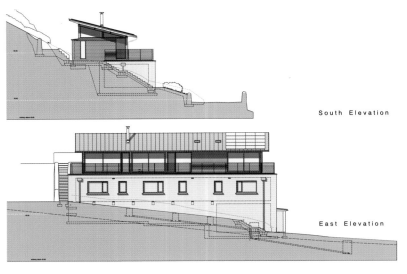

South Elevation

East Elevation

The design uses energy efficient construction and technology: 1. External walls, floor and roof are insulated to a high standard and air infiltration is minimised; 2. Triple glazed windows are with warm edge spacer bars, thermally broken frames and inert gas are filled to achieve a whole window U-value of 0.7W/m²K; 3. Heat pump uses a borehole as the ground source for the underfloor heating and hot water system with a closed combustion wood burning stove as back up; 4. Micro generation of renewable electricity uses roof mounted Photovoltaic Panels and the whole house makes use of heat recovery ventilation system.

本案设计中采用了大量可持续性与节能技术，其中包括：1、建筑外墙、地板以及屋顶均具有很高的隔热性，同时将空气渗透量降到最小；2、三层玻璃窗的边缘使用密封条，窗框具有隔热作用，同时三层玻璃之间有惰性气体填充于内，令窗户U值（传热系数）高达0.7瓦/平方米·开尔文；3、房屋使用地源热泵为地面供暖，并为热水系统提供热能，屋内还增设封闭式燃木火炉作为备用；4、屋顶太阳能光伏板为住宅提供可循环使用的电能，房内还安装有热回收通风系统。

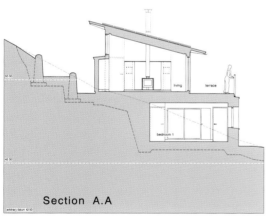

Section A.A

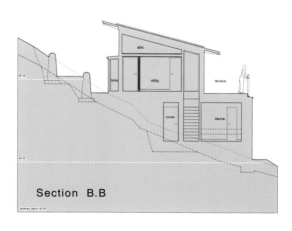

Section B.B

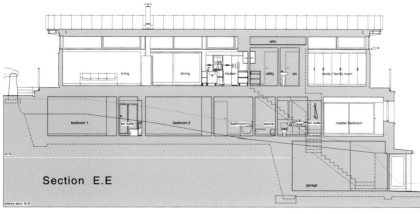

Section E.E

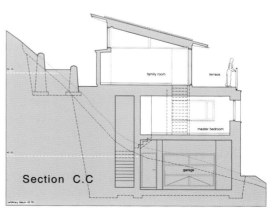

Section C.C

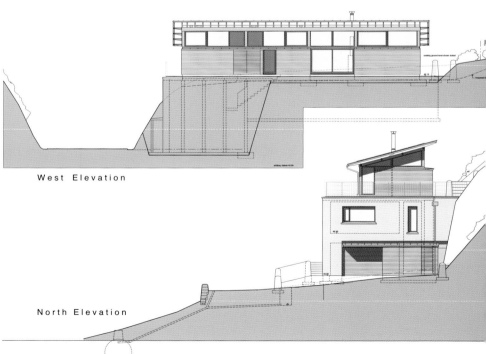

West Elevation

North Elevation

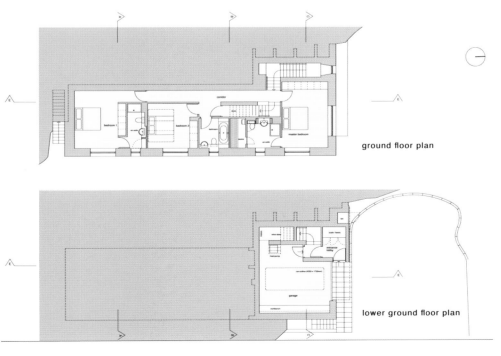

ground floor plan

lower ground floor plan

Elling House in Russia

With an area of 32 square metres, Elling House was completed by Totan Kuzaembaev Architectural Workshop in 2002. It is located in Moscow Region, Russia.

This modest house is one of the first buildings of the new resort established in the early 2000s on the banks of the Pirogovo reservoir as an elite gated community with the complete infrastructure ranging from the golf and yacht clubs to the stables and tennis courts, restaurants and big dwelling houses for purchase or rent.

The house is located near the piers where large boats and yachts are moored, and along the coast there is a number of so-called ellings – small coastal houses of ship-owners, who are "signed up" at Yacht-club "Pirogovo". Those are not countryside cottages or dachas but rather elements of some transitional space, a brief stop before one goes sailing or after one has returned.

由Totan Kuzaembaev Architectural Workshop担纲设计的Elling House位于俄罗斯莫斯科地区，占地面积32平方米，于2002年竣工。

这个简朴的住所乃是21世纪初在Pirogovo水库岸边早期兴建的度假区建筑之一，作为高级封闭式社区建筑，内设完善的基础设施，其中包括高尔夫球场、游艇俱乐部、跑马场、网球场、餐厅及供出售或租赁的大型住宅。

该住所临堤坝而建，坝旁停泊着大型船只和游艇。沿着水岸建有许多被称做"ellings"的建筑，即为Pirogovo水库游艇俱乐部签约船主开发的滨水房。这些住宅不是乡村小屋、别墅，也不同于俄国传统邸宅，而是为人们出海之前或返航之后暂时居住的场所。

Name of Project / 项目名称:
Elling House in Russia
Location / 地点:
Moscow, Russia
Area / 占地面积:
32 m²
Completion Date / 竣工时间:
2002
Architecture / 建筑设计:
Totan Kuzaembaev Architectural Workshop,
Kuzembaev Totan,
Minkevich Dmitry
Landscape / 景观设计:
Totan Kuzaembaev Architectural Workshop
Interior Design / 室内设计:
Totan Kuzaembaev Architectural Workshop
Photography / 摄影:
Ilya Ivanov, Yuri Palmin

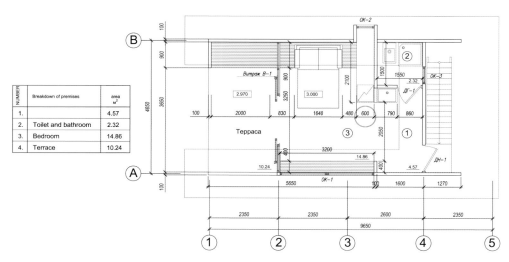

NUMBER	Breakdown of premises	area m²
1.		4.57
2.	Toilet and bathroom	2.32
3.	Bedroom	14.86
4.	Terrace	10.24

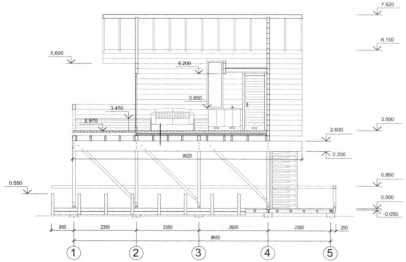

The pile structure installed on the bottom of the bay gives this house its uniqueness. Walls made of glued wooden beams are topped with a gable roof without an attic. Transparent and water-resistant plastic coating acts as the roofing, connecting the interior with the surrounding space.

The first floor is used as a garage for the boat. The drifting vessel is framed by 2 wooden bridges. The second floor is the ascetic residence of the owner-sportsman. It consists of a bedroom and a bathroom. Residential premises are not provided with heating; therefore the house is only used in summer. To save the space, the architect put the bed directly under the arches of the transparent gable roof. It is possible to look at the sky from here, day and night.

The constructions of the house were prefabricated. Placing of the whole volume over the earth level provides minimum intervention in the existing landscape. Therefore, the damage to the environment in the course of construction is extremely insignificant.

This small house integrated into the landscape, open to the water and to the sky, is a place where all the romantic dreams come true.

海湾尽头的桩基结构令本案别有特色。由胶合木梁制成的墙体上是没有阁楼的坡面屋顶。透明的防水塑料涂层覆盖木质屋顶，将室内与周围空间连接起来。

一楼被用做泊船库。漂动的船具以两个木质桥梁为外框。二楼是运动员业主的住处，内部设有一间卧室和一间浴室。室内没有供暖系统，因此这里仅供夏季使用。为节省空间，建筑师直接把床安置在透明斜面屋顶下。这样，无论白天还是夜晚，都可从这里欣赏到天空美景。

该建筑采用预制构件的构造，使地表之上的整个建筑对周围自然景观造成的影响最小化。因此，本案建筑的整个施工过程对环境造成的影响是微不足道的。

这座精巧的小房子与户外自然景观完美融合，面朝绿水与蓝天，似将美好的憧憬一并化为现实。

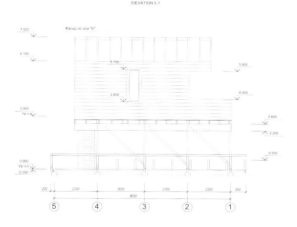

Ellis Residence in Bainbridge Island

Perched high upon Yeomalt Bluff, Ellis Residence enjoys a commanding 180-degree view of Puget Sound and the Seattle skyline. Owners Ed and Joanne Ellis had a special goal in mind when they decided to build a sustainable home in the beautiful surroundings of Bainbridge Island, WA. Their goal was two-fold: they wanted to prove that sustainability can be achieved without compromising a modern aesthetic design and they wanted to motivate others to follow suit.

The home features intimate multi-use spaces that frame exceptional views of Seattle across Puget Sound. The view is best enjoyed from the dining and living room's large expanse of glass windows and doors that open out to a concrete patio. Another great location to capture the vistas beyond is the sunken concrete tub in the master bath and the accessible vegetated roof.

The duration of Ed and Joanne's unique goal encompassed a three year process. This is a small price to pay when your home is a model of sustainability. Ed states it best when he describes the first morning in their home. "When I saw the sunrise with the skyline of Seattle as a backdrop the whole process was all worth it…"

Ellis Residence矗立于Yeomalt山崖之上，与普吉特海湾和西雅图天际线遥相辉映，拥有180度的宽阔视野。当Ed和Joanne Ellis两位业主计划在迷人的华盛顿州班布里奇岛建造一所具有可持续性的住宅时，他们的脑海中就已确定了特殊的目标：首先在不损害现代美学设计的基础上保持可持续性特征；其次要鼓励、引导其他人也接受并效仿这种模式去保护环境。

房内配置了很多精妙的综合用途空间，可以将远处的普吉特海湾和西雅图风光尽收眼底。由混凝土铺制的庭院位于餐厅和起居室的外部，它们之间以巨大的玻璃门窗相隔，使这里成为观赏风景的最佳位置。另一个能够捕获远处美景的极佳位置就是设有浴缸的主浴室和一个可供游览的植物屋顶。

Ed和Joanne的独特想法需要三年的时间来实现，不过这对于成功打造一个可持续性建筑典范来说仅是很小的代价而已。Ed对于他住在这里的第一个清晨是这样描述的："当我看到太阳在西雅图天边升起的那一刻，我就知道这一切都很值得。"

Name of Project / 项目名称:
Ellis Residence in Bainbridge Island, WA, USA
Location / 地点:
Washington, USA
Area / 占地面积:
253 m^2
Completion Date / 竣工时间:
2010
Architecture / 建筑设计:
Coates Design, Inc
Matthew Coates (Principal/Architect),
Justin Helmbrecht,
Bob Miller-Rhees
Landscape / 景观设计:
Outdoor Studio,
Jack Johnson (Landscape Architect)
Interior Design / 室内设计:
Melissa Anderson
Photography / 摄影:
Northernlight Photography,
Roger Turk
Client / 客户:
Ed and Joanne Ellis

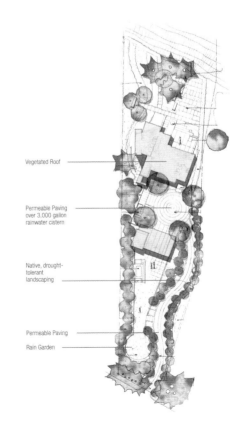

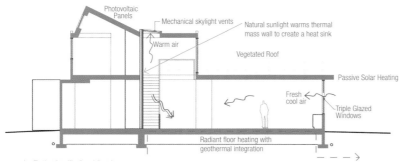

The Ellis's are proud to live in the first LEED© Platinum home outside the city of Seattle. The design limits its impact on the environment with a multitude of sustainable practices. By implementing geothermal, photovoltaic, solar hot water, and advanced heat-recovery technologies, this home has succeeded in reducing its energy consumption by more than 70% compared to a typical home.

The home also makes use of rain water cisterns for irrigation of native landscaping and a vegetated roof. Rather than demolishing the existing structure on the site, the team deconstructed it and effectively diverted around 98% of its material from the landfill.

Ellis很骄傲地说，他们的房子是西雅图城外建筑中第一个获得LEED白金认证的住宅。想要最大限度地减少住宅对环境的影响，需要应用很多可持续性的方式、技术来完成，包括安装地热、光伏发电、太阳能热水器和应用先进的热能恢复技术。该住所相对于同规模的普通建筑来说，节省了超过70%的能耗。

这幢房子还采用水箱积蓄雨水来灌溉户外自然景观和屋顶植被。此外，设计团队并未破坏项目原址上的构造，而是在原有框架基础上进行解构。最终，将近89%的建筑材料都是从废料垃圾中有效回收而来的。

GROUND FLOOR PLAN

GROUND FLOOR CARPORT/ADU
FLOOR PLAN

UPPER LEVEL CARPORT/ADU
FLOOR PLAN

Fab Lab House in Spain

The Fab Lab house was first exhibited at Solar Decathlon Europe, an international competition for universities from all over the world, gearing towards advancing knowledge on industrialized and sustainable homes, with a particular emphasis on high efficiency and energy self-sufficiency.

The Solar House is a new generation Fab Lab home whose goal is to not industrialize but allows any person to manufacture anywhere in the world, from the platform of Fab Labs or fabrication laboratories. The production methodology of the house is founded in a structure fabricated from common materials sourced globally (plywood panels, etc.), and in the use of locally found machinery (laser cutting and/or milling machine). It is definitely a very affordable housing solution, designed with a combination of simple construction, geometric sophistication and technological wealth, both in its creation as an energy system as well as in the active and passive management of the house.

Fab Lab节能房最初在欧洲十项全能太阳能房大赛中展出，这项国际范围太阳能屋赛事面向来自全球各地的所有大学发起，旨在推广工业化可持续性住宅相关的先进技术，着重强调这些住宅的高效性和能源上自给自足的能力。

本案是Fab Lab推出的新一代节能房，其目标并不是生产上的愈发工业化，而是意图让世界上任何一个人，不需要去Fab Lab工作室或是其他制造实验室，只要在其所在地方就可以生产这样的一个节能房。节能房的主要生产方法论基于其所需要的预制结构材料（如胶合板等）和工具（激光切割工具或铣床）在全球各地都极为常见。毫无疑问，该节能房的设计是一个十分经济实用的住房解决方案。它的建造过程简单，综合运用了精确的几何学和先进的技术，同时将一个节能系统与行之有效的主动和被动式管理系统融入到建筑中去。

Name of Project / 项目名称:
Fab Lab
Location / 地点:
Barcelona, Spain
Completion Date / 竣工时间:
2010
Architecture / 建筑设计:
IAAC
Interior Design / 室内设计:
IAAC
Photography / 摄影:
Adrià Goula

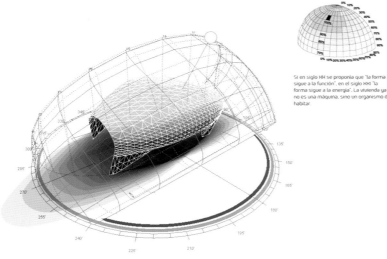

Si en siglo XX se proponía que "la forma sigue a la función", en el siglo XXI "la forma sigue a la energía". La vivienda ya no es una máquina, sino un organismo de habitar.

The selection of wood, not steel, as the basic structural material is deduced from two lines of thought, the first being that a solar house must be reduced from a solar material and the second that the choice of wood leads to structural elements and components which are small, light and manageable. Thus material, scale of building elements and level of technology used should be selected not only for their excellence and functional optimization, but also by its availability and ease of use and maintenance.

Through the utilisation of a global network of production laboratories (Fab Labs), the team begin to promote the idea of using the Internet to make things. The computer, attached to a laser cutter allows people to print a chair, for example, with a wood-cutting machine and then assemble it. The prototype reduces energy expenditure by 25%.

之所以选择木材而非钢铁作为基本构造材料，主要出于两方面的考虑：首先，一所太阳能房必须要采用利于阻隔阳光热能的材料；其二，木质材料的选择会让房屋构造、组件更轻、更小，令建筑过程更加简易。因此，Fab Lab节能房材料的选择、建筑元素的规模以及所用的技术不仅要求有卓越的品质和良好的性能，同时还要容易获取，易于使用和维护。

通过Fab Lab遍布全球的生产实验室的长期实践，设计团队开始推行运用计算机来建造事物的观点。例如，利用附有激光切割机的电脑就可以制造一把椅子，仅需设计出样稿，据此将木头切开，然后重新组装。Fab Lab的这一节能样板房可以将能量损耗减少25%。

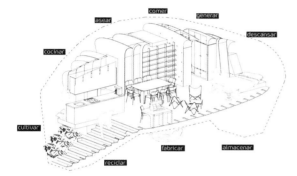
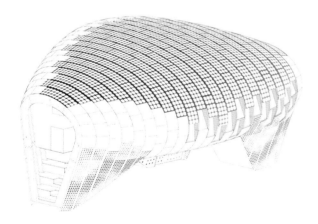

Hillside House in Mill Valley

Nestled in the hills of Mill Valley, California, just across the Golden Gate Bridge from San Francisco, this spectacular custom home has just received certification as the first LEED for Homes Platinum custom home in Marin County, and one of only a handful in Northern California.

Designed by San Francisco-based SB Architects, an international firm well-known for the design of site-sensitive resort and mixed-use projects around the world, and built by well-known green builder McDonald Construction and Development, this project represents a unique approach to the design process.

In a truly collaborative effort, a visionary team of sustainable designers, product manufacturers, local craftsmen and community members worked hand-in-hand to bring this first-of-its-kind project to Marin County. Scott Lee, President and Principal of SB Architects, designed this home as his own residence and, in the process, and produced a definitive statement of what is possible in combining high design with high sustainability.

该住所坐落于加利福尼亚州的米尔谷山丘上，对面是旧金山金门大桥。这一精心打造的住宅刚刚获得了马林郡首个LEED白金认证，它也是加州北部获得该节能认证的少数建筑之一。

本案由享誉国际、位于旧金山的SB建筑公司设计，该公司以世界级度假村和多功能项目设计而闻名。该建造工程由著名的绿色建筑发展公司McDonald负责完成，因此本案成为了建筑史上独一无二的里程碑。

可持续住宅设计师和产品制造商、当地的工匠还有社区成员们的完美合作，为马林郡打造了第一个可持续住宅项目。Scott Lee和SB公司的创作过程也成为历史性的创新，同时使高标准设计和可持续性的结合成为可能。

Name of Project / 项目名称:
The Hillside House
Location / 地点:
California, USA
Area / 占地面积:
465 m²
Completion Date / 竣工时间:
2010
Architecture / 建筑设计:
SB Architects
Landscape / 景观设计:
SB Architects
Interior Design / 室内设计:
Erin Martin Design
Photography / 摄影:
Mariko Reed

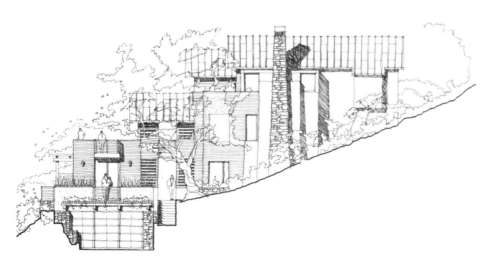

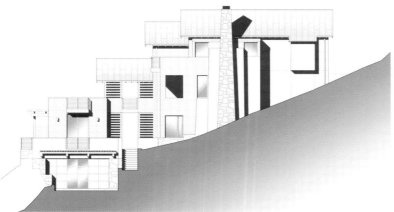

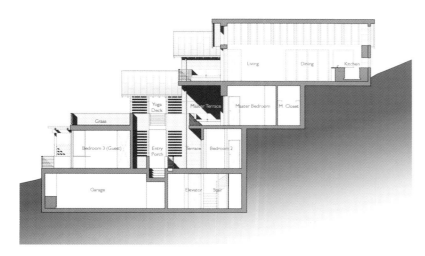

Built on an infill lot close to town, the house is designed to maximize solar orientation for the photovoltaic panels, as well as passive heating and cooling. The surrounding hillside provides the lower floors with natural insulation, solar power supplies electricity and hot water, and radiant floor heating and an innovative air re-circulation system condition the interior. A whole-house automation and lighting system, LED lighting, fleetwood super-insulated doors and windows and indigenous, drought-tolerant landscaping conserve resources.

Local availability, recycled content and sutainable production drove the selection of each material, appliance and detail, including sustainably harvested floors and cabinetry from Plantation Hardwoods and New World Millworks, reclaimed timber and recycled metal roofing. "An important part of minimizing the impact of a project involves selecting products, like Western Red Cedar, that minimize the carbon footprint from manufacture to end use", says Mike McDonald. Design elements crafted locally from reclaimed materials – such as hand-crafted tile from Sausalito-based Heath Ceramics and steelwork from artisan Brian Kennedy – give these project deep roots in the community, making it sustainable from a community standpoint.

Every inch of this LEED Platinum custom home has been designed to maximize its sustainability, in direct response to the site, trees and views. Consequently, this home lives far larger than its actual footprint, but with an impact that is far less.

本案建于比邻市区的一处填埋场地，由于采用了太阳能光伏发电系统，所以房屋尽可能朝向阳面，以期最大限度地吸收阳光为房屋供暖与制冷提供电能。楼下环绕的小山坡成为这幢房子的天然屏障。室内家用电器、地热采暖设施以及新型室内空气循环系统等设施都由太阳能发电系统供电。房内还配置了整套自动化设施、照明系统、LED照明设备、Fleetwood超绝缘门窗，以及源自当地的耐旱景观绿化，这些都在一定程度上起到节约能源的作用。

本案采用的原材料包括来自Hardwoods公司与New World Millworks公司的可持续性环保地板与细木家具，以及可再生木材和可回收使用的金属屋顶。Mike McDonald说："减小破坏影响的重要环节就是选材，像西部红雪松，从制造到最后的使用，都可以有效减少碳排放量。"另有一些由环保材料制成的设计元素，如Sausalito-based Heath Ceramics的手工瓷砖、艺术家Brian Kennedy的钢制作品等，将可持续性原则贯彻于整个项目设计中的各个环节。

这个获得LEED最高等级认证的房屋设计中，每处细节都以可持续性最大化为原则。这一点在项目选址、绿化与周围景观当中都有所体现。最终低碳、节能的房屋所呈现的宽敞效果要比实际空间大得多，而污染却很少。

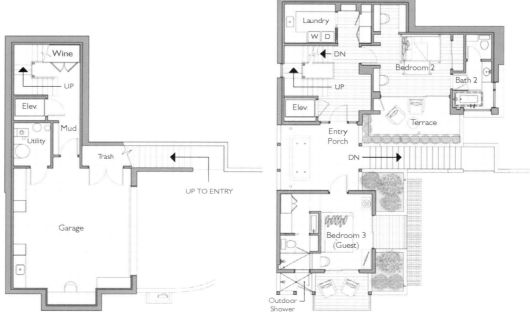

Garage Level

Level 1 (entry)

105

Level 2

Level 3

107

House in Nikaia

Architects design a comfortable and functional home with enough space for each member of the family. The idea is to provide space that is separate but coherent. The family house in Nikaia consists of two cubic forms, a living and a sleeping area, connected by a glass bridge. The two cubic shaped form pavilions provide that kind of intimacy for the family.

In the main object are the living room into which one steps directly from the front entrance, the dining room and the kitchen area. In the upper level is the master bedroom, the laundry room and bathrooms. The secondary cube-shaped pavilion houses the children's bedrooms, one level for each. Through the glass corridor, they have access to the main bedroom and bathrooms.

The floor-to-ceiling height in the main object is 7 metres, with the supporting structure that consists of a steel skeleton reinforced by diagonal elements staying visible. With no conventional internal support walls there is opportunity for more shapely, free form space dividers, reminding of an industrial space that is never designed with rooms in mind.

建筑师为业主打造了一处舒适性与功能性兼备的住宅，足够的室内空间足以令每位家庭成员充分享受生活。建筑师意在打造彼此分离却相互关联的室内空间。这座位于Nikaia地区的住所包括生活区和休息区两个立方体状的小屋，中间由玻璃天桥相连。这两个立方体状的空间成为一家人亲密团聚的场所。

主体建筑包括客厅、餐厅及厨房，进入室内，首先映入眼帘的便是客厅。楼上设置主卧、洗衣室及浴室。双层立方体小屋是儿童卧室，每个孩子各占据一层。透过玻璃走廊，可以看到主卧和浴室。

主体建筑区域从天花到地面间的举架高达7米。支撑结构由对角线元素加固的钢骨架构成，清晰可见。没有传统的内部支撑墙，使更为自由随意的室内布局成为可能，同时这种设计也令人们不禁联想起未经设计的工业空间。

Name of Project / 项目名称:
House in Nikaia
Location / 地点:
Larissa, Greece
Area / 占地面积:
200 m²
Completion Date / 竣工时间:
2008
Architecture / 建筑设计:
Christina Zerva Architects
Interior Design / 室内设计:
Christina Zerva Architects
Photography / 摄影:
Mihajlo Savic
Client / 客户:
Irini and Vasilis Hristodoulias

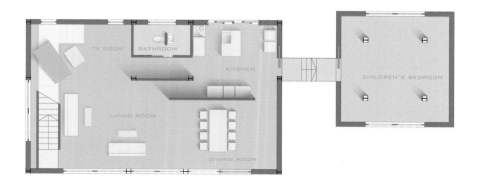

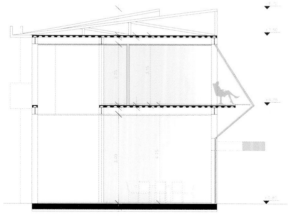

Steel constructions form the skeleton of the object wrapped by prefabricated organic blocks that act as perfect insulation during warm summers and cold winters typical for this region. Interactions of geometrical shapes, huge volumes and transparent surfaces offer a great view in Olympus mountain. Photovoltaic solar panels hidden in the roof top provide sustainable electricity for any use. Recycled and second hand materials have been used for the building. The interior and the exterior is entirely illuminated with LED modules, with low energy consumption.

Architects decided to use Ytong block because of its optimum fire resistance since fires are frequent in Greece. It is also convenient for its thermal insulation. It is ecological and durable. The CO_2 emissions are very low, and, as a result, the environment is less polluted.

Architects chose steel because of its advantages over concrete. It is light and very strong. The main advantage of steel is that it withstands unfavorable weather conditions as well as earthquakes. Steel also has positive environmental impacts for its sustainability, refurbishment, recyclability and reusability characteristics. For those reasons, it is an environmentally friendly building material.

钢结构构成主体建筑区域的骨架。预先设置的有机楼体使本案无论在温暖的夏天还是寒冷的冬季都具有很好的隔温效果。几何形体、巨大体积、透明表面的设计令本案在奥林匹斯山地区一展壮观的景象。隐藏在屋顶的太阳能光电板为家居供应可持续电力能源。可循环利用的材料及二手材料在该建筑中都有所采用。室内外完全利用低能耗的LED灯照明。

建筑师决定选材抗燃性较好的YTONG轻砂加气混凝土块,主要基于希腊火灾频发状况的考虑,且该材料利于保暖。作为生态持久性建筑,其碳排放量极低,因此对保护环境起到积极作用。

建筑师选用钢铁建材的原因在于,它与混凝土相比更具有轻巧、坚固的性能优势。同时,它可以禁得起恶劣气候环境的考验,并具有良好的抗震效果,对保护环境也会起到积极作用。此外,可翻修、可持续性、再循环性和再利用性也是这种建材的特点所在。由此,钢铁可以被称做环境友好型建筑材料。

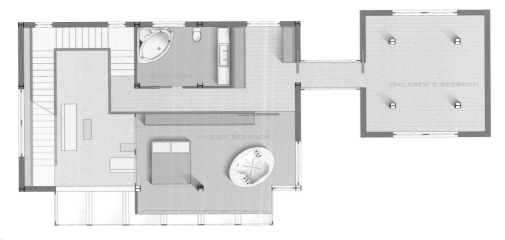

Iseami House in Costa Rica

This secluded project site is located at 30 km from the closest town, Puerto Jimenez, therefore, the house doesn't have any public service supply as electricity or water, and this condition obligates the institute to invest on a 100% self-sufficient house. During the design process, decisions were takenin order to bring the water and energy supply. The existence of a water concession on the protected forest inside the property, allows the project to consider a natural potable water source for its use, then the water volume is utilized in order to produce energy with 2 low impact hydroelectric generators that generate a total of 800 kWh. Furthermore, Casa Iseami becomes a solar power plant with a production capability of 10,800 kWh, this because the roof has been designed in order to have the orientation and position to maximize its production. The roof material has great structural capacities, so the access to the roof for maintenance is possible (The use of a solar hot water tank provides comfort to the users). The hybrid energy system provides with success enough energy to the project, the caretaker'sand maintenace housing and the existing lodge for the participants of the institute.

The materials for the house were selected following the results given by an investigation process done to the existing project near the site. This way the architects took the results and then decide which were the best materials to use in order to create a low maintenance project. All the proposed materials were tested in order to evaluate its behavior on the difficult environment that the Osa peninsula has because of its humidity, high percentage of rain, mould, fungi and its flora and fauna.

Name of Project / 项目名称:
Iseami House
Location / 地点:
Puntarenas, Costa Rica
Area / 占地面积:
482 m²
Completion Date / 竣工时间:
2010
Architecture / 建筑设计:
ROBLESARQ Architecture Studio
Landscape / 景观设计:
ROBLESARQ Architecture Studio
Interior Design / 室内设计:
ROBLESARQ Architecture Studio
Photography / 摄影:
Sergio Pucci

本案所在位置十分隐蔽，据它最近的城镇Puerto Jimenez也有3万米的路途，这里无法得到任何公共服务供给，不通水电。如此的形势要求建筑团队必须去打造一个100%自给自足的家庭住所。在设计过程中，建筑师提出许多可以给房子引入水源和能源的设计。由于住宅位于一片受保护的森林内，大量水资源的存在允许建筑师将其作为饮用水的来源，同时利用两个低强度水力发电机发电，为住宅提供能源，两个发电机共可发电800千瓦时。Iseami住宅还是一个拥有10 800千瓦时生产能力的小型太阳能发电设施，其精心设计的屋顶方位恰到好处，使其所产生的太阳电能达到最大。屋顶所用材料具有极强的承载力，使得在屋顶上进行维护成为可能（太阳能热水槽更为使用者带来舒适）。混合的能源系统可为该住所带来足够的能源供给。

对该住所附近一个已存项目的调查结果，为本案材料的选择提供了有力的依据。通过参考这次的调查结果，建筑师明确了应该使用哪些材料才能打造一个易于维修的建筑项目。这里湿度高、降水量大、易发霉和滋生菌类植物，及其周围遍布多种动植物群等特点，使奥萨半岛的环境相对严酷，因此，所有材料都必须经过测试，评估其是否适合如此严酷的环境。

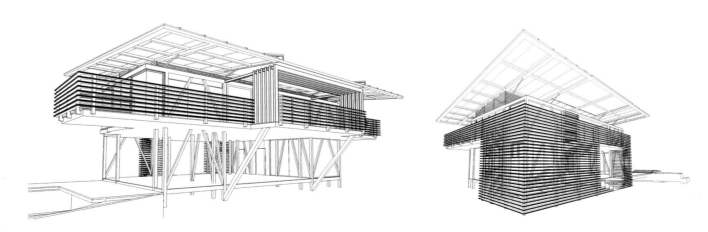

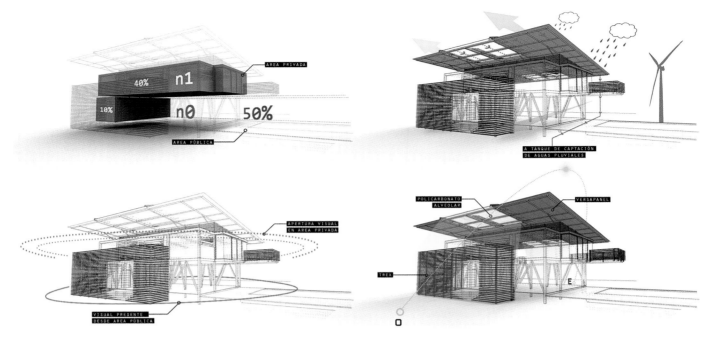

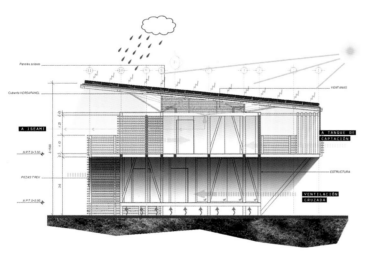

The passive design strategies have been successfully implemented to handle the sun exposure, relative humidity, natural illumination and ventilation inside the bioclimatic considerations of the project. Relative humidity is managed to create a healthy and comfortable space. The strategies used are, elevating the house 1 metre above the ground (water permeability in the ground will be possible), ideal orientation (southeast-northwest) for cross ventilation and the installation of two polycarbonate skylights that provide the control of sunlight exposure in the interiors of the house, preventing UV to damage the furniture and Indoor environmental quality for its occupants. Large overhangs create shadows all day long in order to control the interior temperatures and natural illumination together with the thermal capacities and white colour of the envelope.

The use of recycled plastic louvers (Trex) creates a second envelope that surrounds the house in order to maximize the temperature control and rain exposure. On the other hand, the white colour in the totality of Iseami has the goal to be able to visualize any insect or mould creation inside the house, frame and create a contrast with nature, as well as maximize natural light and solar reflectance index (SRI 100). The roof design allows the occupants to have an integration with the mountains around with an "open to the sky" concept, and in this way, cross ventilation, daylight and views are present in every single space around the house.

出于对生物气候性因素的考虑，建筑师成功地将被动式设计策略应用在Iseami住宅中，解决日照强度、相对湿度、自然照明以及空气流通等问题。这些被动式设计策略包括：将房子抬高于地面一米，以增强地面土壤的透水性；房子呈完美的东南－西北朝向，利于形成对流通风；同时，屋顶还装有两个聚碳酸

酯天窗,以控制室内日光强度,并防止紫外线损害室内家具和影响居住者所处的室内环境质量。大型的悬臂结构形成的阴影可以持续一整天,有效地控制了室内温度和日光照明。

可循环的塑料百叶窗的使用为房子创造了第二层防护,最大限度地控制住宅温度并防止雨水侵袭。同时,Iseami住宅几乎是全白的,这种设计不仅可以令室内的昆虫和发霉处一目了然,同其所处的自然环境形成鲜明对比,同时也可让房子尽可能接触自然光,将其太阳能反射指数最大化(SRI指数高达100)。住宅"面朝天空"的屋顶设计令居住者仿佛同自然融为一体,如此的设计使对流通风、日照和美景存在于房子的每个角落。

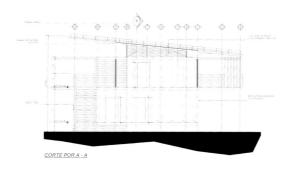

CORTE POR A - A

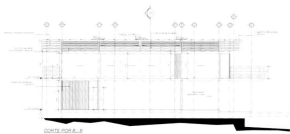

CORTE POR B - B

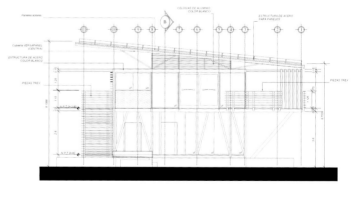
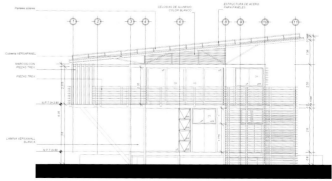

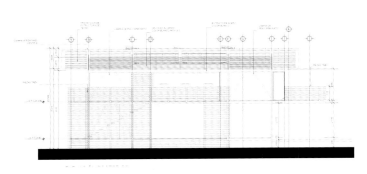
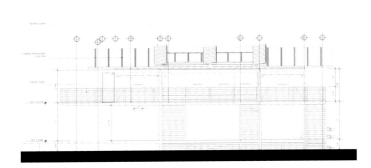

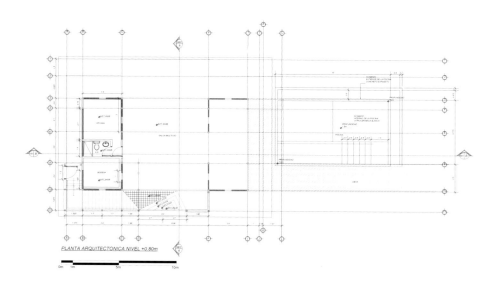

PLANTA ARQUITECTÓNICA NIVEL +0.80m

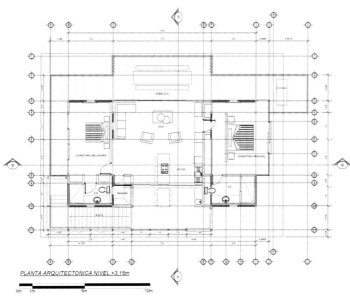

PLANTA ARQUITECTÓNICA NIVEL +3.15m

Knowles Residence in Seattle

For the architectural remodeling of Knowles Residence, Coates Design opted for a profound lifestyle makeover. The aim of their design was to breathe a new life into a mundane dwelling, transforming it into an architectural living space with soul and character.

The remodel of the Knowles Residence transformed a small and uninviting home to a place with heart, substance and style for a family of four. The existing house had no central place for occupants to naturally congregate. Occupants complained that the house was uncomfortable and confining. The design team set out to remedy this with great momentum and ultimately, success.

The design created a dramatic entry into a living space where inhabitants feel inspired to relax, rejuvenate, gather and linger. High ceilings and abundance of natural light create a feeling of space, and at night, the home glows like a lantern. Asian influences throughout the design process create a "Zen" of serenity and tranquility.

Coates Design将Knowles Residence重新翻修，为其选择了寓意深刻的家居风格。设计目的就是要为这处平凡的住所带来新的生命，给这个生活空间带来灵魂和个性。

Knowles Residence的翻修对于这个四口之家来说是一次全新的体验，原本十分狭小、毫无吸引力的住宅被改造成温馨实用、独具风格的家。改造前的住宅没有聚会的地方，购买者还抱怨房屋缺乏舒适性，且很有局限性。设计团队就这一问题从根本上找到了十分有效的解决方案。

设计师们在通往起居室的地方建造了一个引人注目的入口，这样，住户就可以在这里任意的休闲、放松、聚会和休息。高高的天花板和充足的自然光线营造出一种独特的氛围。在夜晚，这里看起来就像灯笼一样，耀眼夺目。富有亚洲特色的设计贯穿于整个住宅之中，营造出"禅"的宁静与祥和。

Name of Project / 项目名称:
Knowles Residence in Seattle, WA, USA
Location / 地点:
Washington, USA
Area / 占地面积:
229 m²
Completion Date / 竣工时间:
2008
Architecture / 建筑设计:
Coates Design, Inc,
Matthew Coates (Principal/Architect),
Bob Miller-Rhees,
Justin Helmbrecht
Landscape / 景观设计:
Outdoor Studio,
Jack Johnson (Landscape Architect)
Interior Design / 室内设计:
Coates Design, Inc,
Matthew Coates (Principal/Architect),
Bob Miller-Rhees,
Justin Helmbrecht
Photography / 摄影:
Michael Cole
Client / 客户:
Eric and Judith Knowles

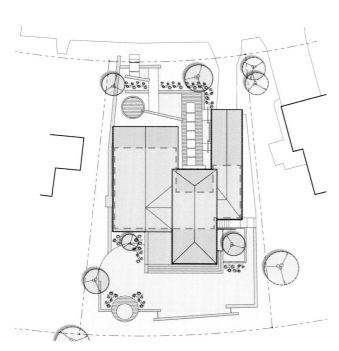

The clients were open to dramatic change, and the design team embraced the opportunity to create a hearth for the home, walls were removed and ceilings were raised to create an open concept plan. A radiant floor heating system was installed, creating efficient use of energy and reduction of utility bills. The home was transformed into a living space with 4-metre ceilings and floor to ceiling windows, creating a sense of spaciousness and light during the day which gracefully transforms into a glowing lantern in the evening. The Knowles family now has a home where they feel truly comfortable.

Because of their budget constraints, the Knowles wanted to focus their budget the sustainable design measures that yielded direct results. Some of the features and materials include: radiant-floor heating HVAC retrofit, zero VOC paints and sealers, high-efficiency induction cook-top, recycled glass tile, recycled rubber flooring, reconstituted concrete countertops, FSC certified eucalyptus flooring, high-efficiency lighting design, drought tolerant landscaping, and "iron wood" plank decking.

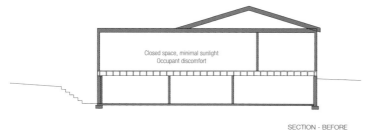

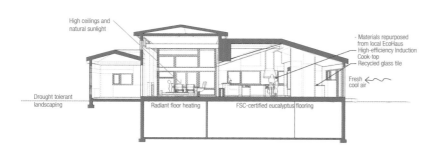

设计团队抓住了客户愿意接受突破性改变这一机会，将住宅的墙壁改造为一个壁炉，同时升高了天花板，这样就体现出一个开放性的概念创意。一个地热供暖系统也被安装于此，有效地利用了能源，减少了水电费用的开支。改造后的居住空间有着4米高的天花板和巨大的落地窗，在视觉上更显开阔。白天日光闪耀，到了晚上，又变成了一个发光的灯笼。现在，Knowles一家真正拥有了一个舒适温馨的居住环境。

由于预算的限制，Knowles想将主要预算都用在可持续性设计上去。这种应用和材料主要包括：地热供暖、空调改造、零VOC（挥发性有机化合物）涂料和密封材料、高效能电磁感应厨房用具、可循环的玻璃瓷砖、再生橡胶地板、再造混凝土台面、经过FSC认证的桉木地板、高效能照明设计、耐寒景观绿化以及"铁木"装饰板。

FLOOR PLAN - BEFORE

FLOOR PLAN - AFTER

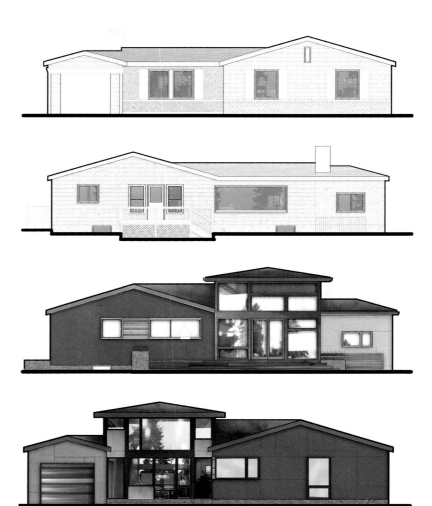

127

Kona Residence in USA

Completed in 2010, this stunning 743 square metres residence by Belzberg Architects is one of the finest architecture. Located in Kona, Hawaii, the property is situated between volcanic mountain ranges to the east and ocean horizons to the west. Nestled between cooled lava flows, the Kona residence situates its axis not with the linearity of the property, but rather with the axiality of predominant views available to the site. Within the dichotomy of natural elements and a geometric hardscape, the residence integrates both the surrounding views of volcanic mountain ranges to the east and ocean horizons westward.

The program is arranged as a series of pods distributed throughout the property, each having its own unique features and view opportunities. The pods are programmatically assigned as two sleeping pods with the common areas, media room, master suite and main living space. An exterior gallery corridor becomes the organizational and focal feature for the entire house, connecting the two pods along a central axis.

由Belzberg Architects担纲设计的惊艳别墅之作于2010年竣工，占地面积为743平方米。本案位于夏威夷柯纳地区，向东可抵火山山脉，以西则遥望海岸线，该住所就建于周围冷却的的熔岩流之间。本案并没有完全追随原有的线性构造，独特的位置安排可以将户外美景尽收眼底。受到自然元素和几何硬性景观双重因素的影响，别墅融合了东部火山山脉与西部海洋的风景。

该别墅内部被间隔成一系列小房间，每一间都独具特色，并且可供居住者欣赏到迷人的风景。设计师为两个休息间设计了公共区、媒体室、主套房及主客厅区。位于中央位置的户外走廊沿中轴线将两侧房间连接起来，成为整栋建筑贯穿一体的重要组成部分，也是本案设计的亮点所在。

Name of Project / 项目名称:
Kona Residence
Location / 地点:
Hawaii, USA
Area / 占地面积:
743 m²
Completion Date / 竣工时间:
2010
Architecture / 建筑设计:
Belzberg Architects
Landscape / 景观设计:
Belt Collins Hawaii
Interior Design / 室内设计:
MLK Studio
Photography / 摄影:
Belzberg Architects, Benny Chan

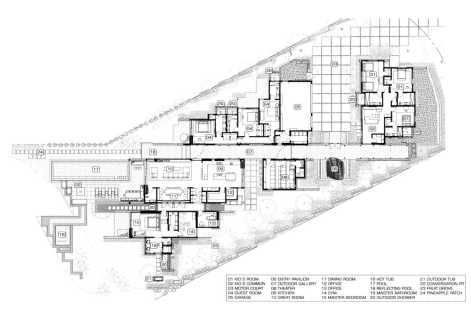

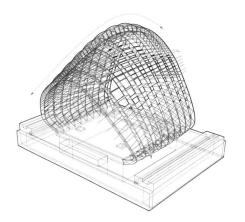
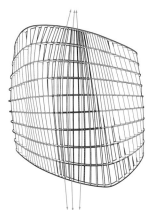
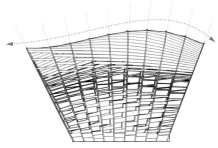

The 3D modeling and digital fabrication through CNC machinery was incorporated to further enhance traditional elements in a contemporary arrangement. Local basket weaving culture is the inspiration for the entry pavilion which reenacts the traditional gift upon arrival ceremony. Various digitally sculpted wood ceilings and screens, throughout the house, continue the abstract approach to traditional Hawaiian wood carving further infusing traditional elements into the contemporary arrangement.

To help maintain the environmental sensitivity of the house, two separate arrays of roof mounted photovoltaic panels offset the residence energy usage while the choice of darker lava stone help heat the pool water via solar radiation. Rain water collection and redirection to three drywells that replenish the aquifer are implemented throughout the property. Reclaimed teak timber from old barns and train tracks are recycled for the exterior of the home. Coupled with stacked and cut lava rock, the two materials form a historically driven medium embedded in Hawaiian tradition.

通过数控机械技术而研制开发的三维建模和数字制造产品在当代技术中进一步强化了传统元素。入口处的设计深受当地编篮文化的启发，将传统文化融入其中，扮演迎送宾客的角色。从天花板到墙面，数控切割木材似乎无处不在，将抽象的夏威夷传统木刻技术灌输到现代文化之中。

为保证房屋的环保性，屋顶位置装有两个独立的太阳能光电板，为别墅提供能源供给。深色熔岩石通过太阳能辐射加热池水。雨水收集系统将雨水引向三口干井，持续为蓄水层补充水分。从旧谷仓和火车轨道回收而来的再生柚木材被用于室外装饰中，加上堆叠修葺的熔岩石，两种材料的应用形成夏威夷传统建筑史上一次革新的推动。

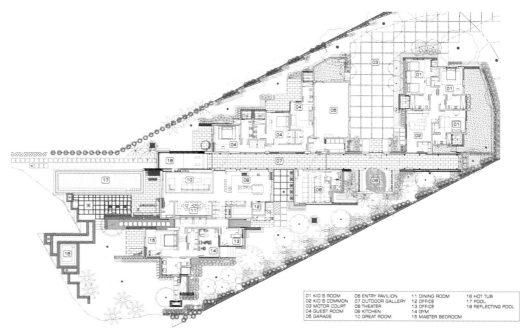

01 KID'S ROOM	06 ENTRY PAVILION	11 DINING ROOM	16 HOT TUB
02 KID'S COMMON	07 OUTDOOR GALLERY	12 OFFICE	17 POOL
03 MOTOR COURT	08 THEATER	13 OFFICE	18 REFLECTING POOL
04 GUEST ROOM	09 KITCHEN	14 GYM	
05 GARAGE	10 GREAT ROOM	15 MASTER BEDROOM	

Kyneton House in Australia

This single-story house is located on a leafy well established street in Kyneton, Victoria, Australia. Whereas the typical suburban model of a distinct front yard and back yard is the norm in the area, this design focuses the house to a generous north facing outdoor room.

The house's dramatic entry steps up from the gravel driveway and has recessive garaging and water tanks to either side. This entry steps up to a concrete path which runs through the house as an axis: with the active areas of the living, dining, kitchen on one side opening out onto the sunny, north-facing deck and garden. On the other side are the quieter areas of the house, for sleeping, the study or the laundry, each opening onto a courtyard.

Stretching east to west across the site, the dynamic double skillion roofs provide ample northern light into every room in the house. A central spine between the two roofs splits the house into public and private functions and visually connects the entrance of the home to the sculptural forms of the rear garden.

这座只有一层楼高的住宅位于澳大利亚维多利亚州基内顿的大街上。典型的郊区住宅楼一般都有明显的前院和后院，而这一住宅在设计上却强调了面朝北侧的户外家居部分。

入口处十分引人注目，它从房前的碎石车道延伸出来，道路两边是隐蔽的车库和水箱。一进入口，就能看到混凝土过道，它是一条贯穿整座住宅的轴线，轴线一侧是面朝太阳的居住区、餐厅、厨房和朝北的露天平台、花园。另一侧是住所内较为安静的区域，可以用做卧房、书房或洗衣房，这里的每一处都可以通往庭院。

富有动感的双层坡形屋顶从东到西，贯穿整个房屋，使住宅里的每一个房间都有充足的光源。屋顶的中间轴线将住宅分为公共区和私密区两部分，这一设计将住宅入口和它的后花园连接起来。

Name of Project / 项目名称:
Kyneton House
Location / 地点:
Victoria, Australia
Completion Date / 竣工时间:
2010
Architecture / 建筑设计:
Marcus O'Reilly Architects
Landscape / 景观设计:
Marcus O'Reilly Architects
Interior Design / 室内设计:
Marcus O'Reilly Architects
Photography / 摄影:
Dianna Snape

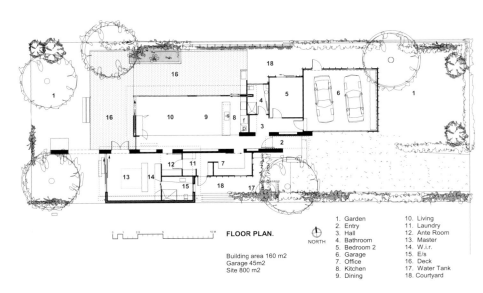

FLOOR PLAN.

Building area 160 m2
Garage 45m2
Site 800 m2

1. Garden
2. Entry
3. Hall
4. Bathroom
5. Bedroom 2
6. Garage
7. Office
8. Kitchen
9. Dining
10. Living
11. Laundry
12. Ante Room
13. Master
14. W.i.r.
15. E/s
16. Deck
17. Water Tank
18. Courtyard

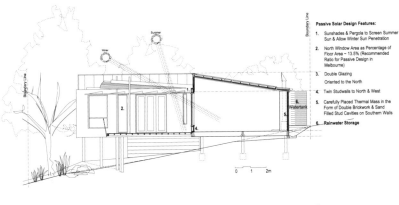
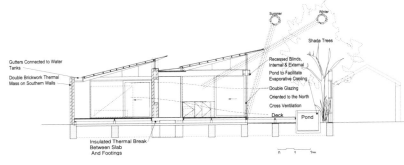

The carefully tapered wide eaves let the winter sun onto the ample provision of thermal mass in the floor and southern walls, and the double glazing full of sealed air gap under the floor all work so well that top up heating is usually only required for an hour or so in the morning during winter. Given that Kyneton is freezing in winter, this is quite something!

In reverse fashion in summer, with the sun controlled, the thermal mass can keep the house so cool that there is no need for air conditioning.

None of the rooms need lighting till dark, even on the greyest of days. And the light gives a feeling of space that belies the house's relatively modest footprint – the 160 square metres garaging.

Completing the ESD package are solar power panels providing power connected to the grid, and solar hot water, and panels heating for hydronic heating, including front concrete panel of the kitchen bench of polished concrete which, like all the heating panels, are integrated into the building fabric.

这种精心设计的锥形宽屋檐可以使房屋在冬天里也有充足的阳光，保持室内地面和南面墙壁的热量。地板下面装有密封气体的双层玻璃，这样可使Kyneton住宅即使在冬天早上也只需加热一小时就可以保证温度，因为基内顿的冬天也不容易结冰。

夏天则刚好相反，在控制阳光的同时，其隔热作用能够保持室内凉爽，从而不需要使用空调。

不到天黑的时候，房间不需要灯光，即使在天气最阴的时候，也是如此。灯光的设计营造出了一种空间感，掩饰了这所住宅算上车库也仅有160平方米的局限性。

太阳能板配有防静电包装，它为电网、太阳能热水器、热水循环加热板提供能量，此外，抛光混凝土厨房工作台前面的混凝土板也像其他加热板一样，被融合到建筑物中去。

Kuhlhaus 02 in USA

Kuhlhaus 02 is single family residence designed with minimalist house style. This house located in the Gaslight District of Manhattan Beach is one in a series of homes commissioned by Kuhlhaus Development, LLC and produced by the award-winning Los Angeles based Design/Build firm Lean Arch, Inc. The house design is committed to responsible development, redefining the modern home in response to concerns regarding energy consumption, land-use and the environment.

This solar powered home features 4 bedrooms with four bathrooms, powder room, kitchen by Valcucine, open living and dining areas, family room, recreation room with a wet bar, walk-in wine cellar, and laundry room. The open plan and large glass sliding pocket doors create both visual and physical continuity between the interior and exterior. Outdoor patio and deck space adjacent to the kitchen, and the living and family rooms, and the master bedroom enhance the quality of each space and allow for flexibility in their use. A sophisticated colour scheme has been designed to enliven the industrial Grey. Kuhaus 02 is the second model for the Kuhlhaus Development Company, and designed to expand on the original, literally. This one, 380 square metres with the garage, by contrast with the first model, is huge.

Kuhlhaus 02是一栋极简风格的独立式住宅，位于曼哈顿海滩的Gaslight区，是Kuhlhaus发展有限公司开发的一系列住宅中的一个。Kuhlhaus 02由备受赞誉的洛杉矶设计团队Lean Arch建筑设计事务所担纲设计。本案设计旨在重新定义现代住宅，充分考虑建筑能耗、土地利用以及环境污染等问题。

Kuhlhaus 02住宅以太阳能作为其主要的能量来源，空间包含四个卧室及浴室、盥洗室、Valcucine厨房、开放的会客区和用餐区、家人休息室、一间拥有小吧台的娱乐室、一个步入式酒窖和洗衣间。住宅采用开放式的设计，巨大的玻璃滑动式折叠门令室内和室外空间无论是在视觉上还是物理上，都是一个协调、连续的整体。室外露台靠近厨房、客厅、家庭娱乐室以及主卧房，不仅提高了室内的环境质量，同时还增强了每个空间使用上的灵活性。为本案特别设计的颜色主题精致考究，使得这个以工业灰色为主色调的住宅异常生动。Kuhaus 02住宅是Lean建筑事务所为Kuhlhaus设计的第二个样板住宅，并在第一个基础上加以扩大。包括车库在内，Kuhlhaus 02的面积达到380平方米。

Name of Project / 项目名称:
Kuhlhaus 02 in USA
Location / 地点:
Manhattan Beach, USA
Area / 占地面积:
380 m²
Completion Date / 竣工时间:
2008
Architecture / 建筑设计:
Lean Arch, Inc. Design + Construction
Landscape / 景观设计:
Jones & Potik, Inc
Photography / 摄影:
Claudio Santini
Client / 客户:
Kuhlhaus Development Company, LLC

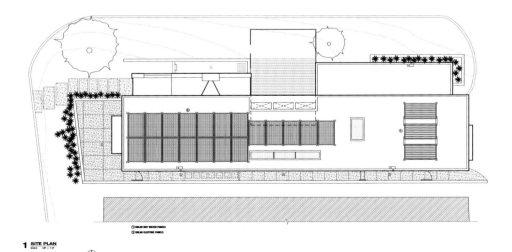

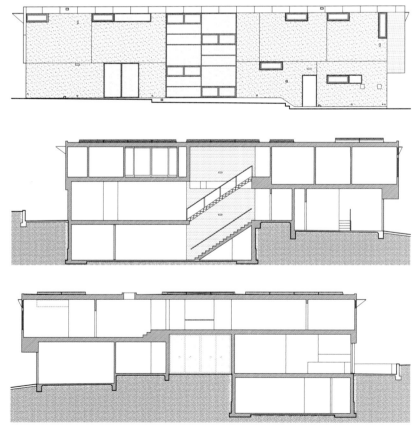

James Meyer of Lean Arch wants fashion homes to show that it is possible to enjoy a sophisticated urban beach lifestyle in houses that are sustainable and responsible with regards to land and energy use. A perforated metal stair, metal panel siding, radiant floor heating and a 4.3 KW array of solar panels on the roof are a nod to "industrial chic". Rooftop solar panels generate all the electricity needed. Water is heated by the sun and circulated through radiant flooring embedded in the industrial concrete floors. As with Kuhlhaus 01, Meyer looks to commercial mass production technology like SIPs (Structural Insulated Panels) like these modules on the exterior. This kind of construction enables buildings to be put together faster, cheaper and less wastefully, during construction.

来自Lean Arch公司的James Meyer表示，建筑团队的主要目的是改变传统的住宅形式，让主人可以在这里享受到郊外海滨的精致生活方式，同时，住宅还需对其所使用的土地以及能量消耗负责。穿孔金属楼梯、金属护墙板、地热采暖系统以及屋顶上功率为4.3千瓦的太阳能板，都完美地诠释了"工业设计中的别致之美"。屋顶上的太阳能板可提供住宅需要的所有电能：首先利用太阳能将水加热，然后流通到镶嵌于工业混凝土地板中的地热循环系统。同第一代Kuhlhaus 01一样，Meyer关注适用于工业大批量生产的技术，例如，应用在建筑外观上的结构绝缘板。正是出于这样的考虑，Kuhlhaus 02住宅的建造过程更快速、成本更低廉、造成的浪费更少。

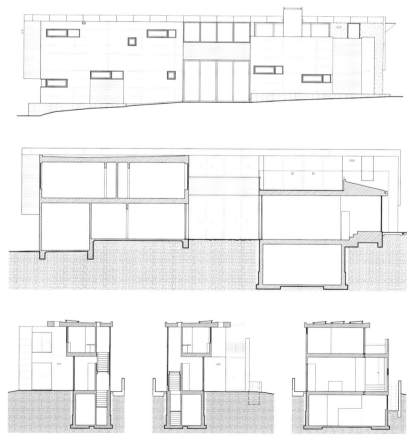

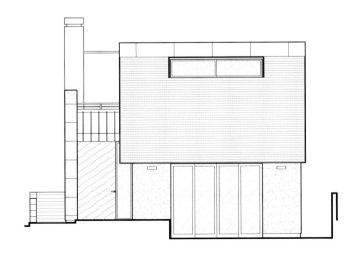

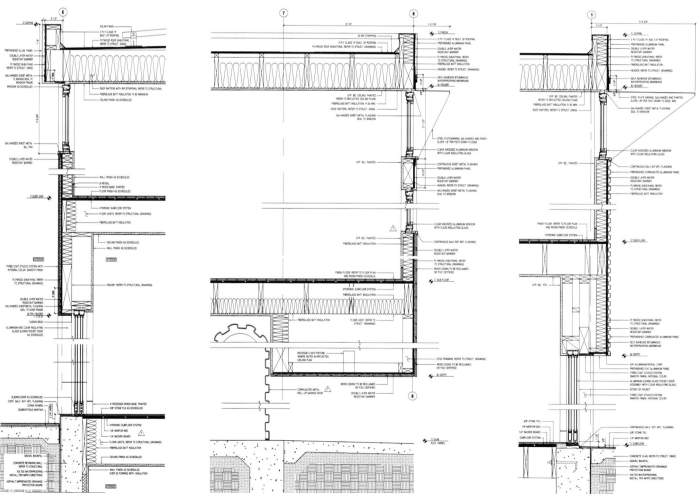

IN GREEN! RESIDENT ARCHITECTURE • HOUSE

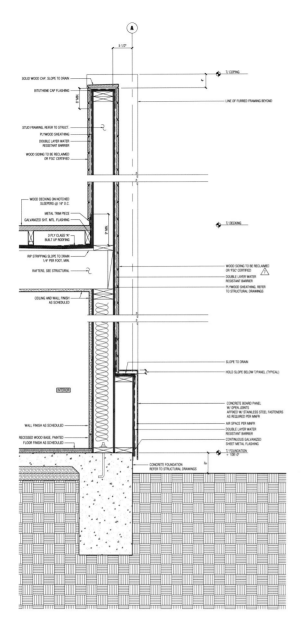

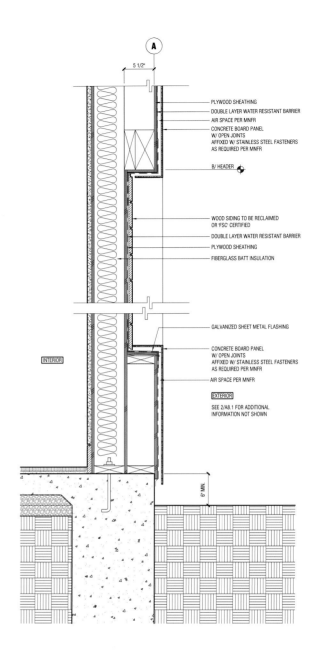

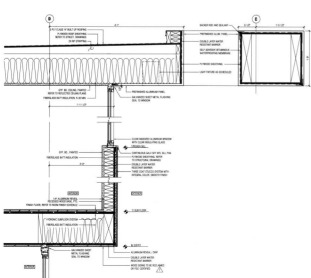

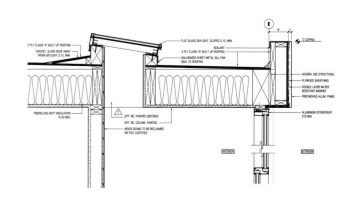

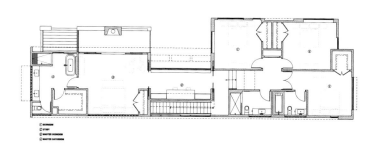

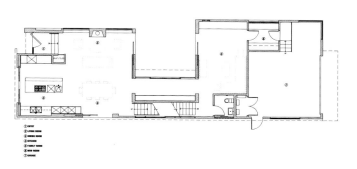

MC1 Residence in Costa Rica

MC1 Residence is located at Manuel Antonio rainforest in central pacific of Costa Rica. For the construction of this Project, it wasn't handed to cut any tree, as the property had a clear area, which was utilized for the construction of the Project. This Project follows a tropical contemporary style and was designed with the main concept of "Integration with landscape and nature with the minimization of environmental impact following bioclimatic and sustainable strategies".

In the Project MC1, the use of natural resources was done responsibly to lower the direct impact, integrating and using the climate, reducing the water and energy consumption, and furthermore, durable materials were chosen, having always in mind the chance of recycling and reusing them, mainly when a possible future extraction is taken into consideration in a long term.

The design of the project resembles the vernacular architecture of the area inherited by the United Fruit Company in the 1900's, which has as a main characteristic the use of bioclimatic buildings and have influenced strongly the architecture in the area. As main characteristics of this architecture, the buildings use the winds and natural lights to create the different ambiences and temperatures inside the house, at the same time, they are lifted from the ground to avoid the humidity of coming inside the house, mainly when the cross ventilation is achieved by the orientation. The use of corridors with long overhangs wrapping the house is something pretty common in this architecture to integrate successfully the interior.

Name of Project / 项目名称:
MC1 Residence
Location / 地点:
Manuel Antonio, Costa Rica
Area / 占地面积:
441 m²
Completion Date / 竣工时间:
2008
Architecture / 建筑设计:
ROBLESARQ Architecture Studio
Landscape / 景观设计:
ROBLESARQ Architecture Studio
Interior Design / 室内设计:
ROBLESARQ Architecture Studio
Photography / 摄影:
MC

MC1住宅坐落于哥斯达黎加中部的曼努埃尔安东尼奥德热带雨林内。由于这里原本就是一块空地，因此项目的实施十分顺利，在整个建造过程中没有一棵树木被砍伐。住宅的设计遵循当代热带建筑风格，设计重点旨在通过生物气候学和可持续发展的方法，将景观和自然完美融合，并把对环境可能造成的影响降到最低。

在MC1住宅的建造过程中，所有自然资源的使用都必须本着对环境负责的原则，避免对环境造成最直接的冲击，建筑师还充分地整合、利用当地气候特点，减少了水资源和能量的消耗，同时，特别选择了持久耐用的材料，以便日后回收再利用，因为只有将未来的发展铭记在心才是长久之计。

本案的设计同当地United Fruit公司始建于20世纪90年代的乡土建筑类似，这类建筑充分利用生物气候，对当地建筑产生巨大影响。其主要的建筑特色包括：有效利用风和自然光，为室内空间打造不同的环境氛围和温度；同时，这些建筑物大多是脱离地面的，以避免室内过分潮湿，而对流通风的取得则主要由建筑物的朝向所决定。

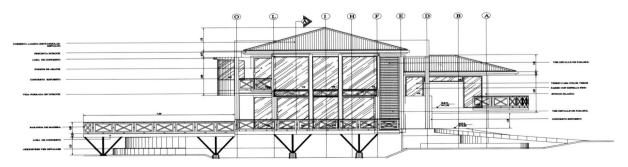

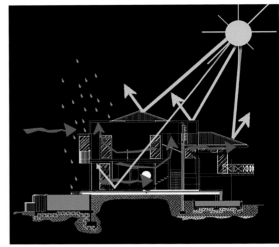

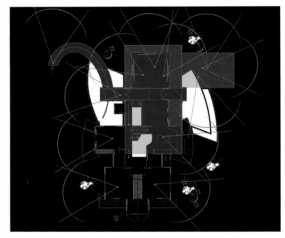
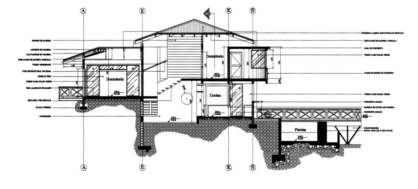
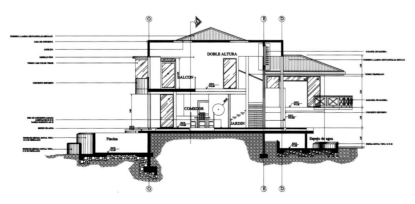

A creation of erosion and sedimentation control plan was created during the design phase and implemented during the construction. The house site is a previously developer land, which was the only area in the property without any tree.

The orientation of the house was chosen to minimize the environmental impact following the natural topography of the property to avoid a high volume of dirt movement and to benefit the house temperatures with the use of crossed ventilation. Native plants have been planted in the surroundings and open interior gardens to create natural habitats and biological corridors, for native species as a high diversity of birds and monkeys.

The parking design promotes the infiltration and reduces the heat islands. The pathway to the house is lifted from the ground to promote infiltration and reduction of dirt movement volume. The 50% of the house is lifted from the ground to benefit the internal temperatures, reducing the humidity inside the house and promoting the rainwater infiltration. A store water collection system was implemented in the house, collecting the rainwater from the roofs and storing it in a captation tank. This water is used for the water mirrors, pool

and internal and external landscape irrigation. Water-efficient appliances and toilets were installed in the house. The use of potable water has been reduced to 50%.

在MC1住宅的设计阶段,建筑师创造性地提出了一个侵蚀和沉积控制计划,这一计划在建设阶段被付诸现实。本案的所在地原是一块待开发的土地,这里是该区域唯一一块没有任何植被的地块。

为使建设工程对环境的影响降到最小,房子的朝向顺应地块的原始地形,既避免了大面积的土地移动,同时利于形成对流通风,调节室内温度。建筑周围种满了本地生植被,室内还设有开放式花园,为本地多样性的物种(比如鸟类和猴子)创造自然栖息地和生物长廊。

停车场的设计渗透性极高,有效地缓解了热岛效应。通向住宅的小路被抬离地面,促进了建筑的渗透性,同时最大程度地减少了土的转移。整个房子有一半的空间是被抬离地面的,这样的设计不仅可以调节室内温度,减少湿气,同时可以促进雨水的渗透。房子内安装了蓄水采集系统,用来收集屋顶雨水并将其储存在集水箱中。采集的雨水将被重新利用在水镜、水池以及室内、外绿地的灌溉中。室内装有各类节水电器和卫浴器具,饮用水的使用量缩减为正常使用量的一半。

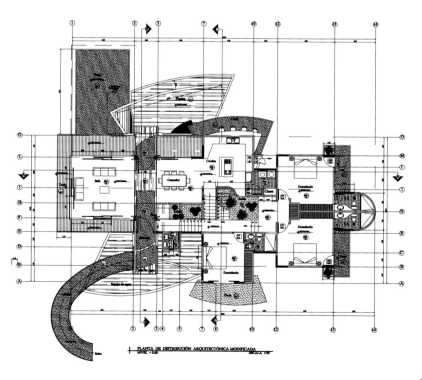

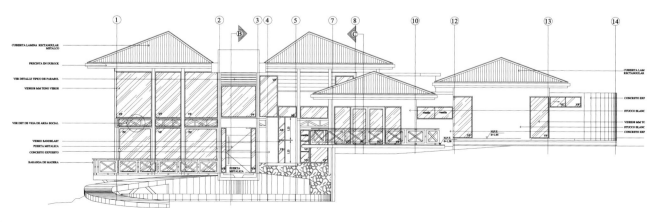

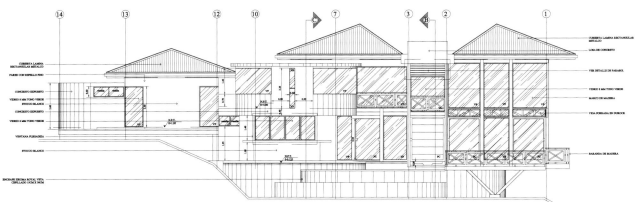

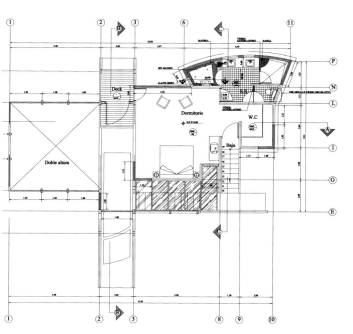

Mosman Green in Australia

This project involved alterations and additions an existing late Victorian/ early Federation double-storey house in Mosman. Previous renovations added to the rear of the home creating rooms that relied on others for light and ventilation and a first floor bedroom of poor design and construction.

Conceptually, the design connects the new living spaces with the rear yard while reinforcing the retained original rooms, which are grouped around a central entry and hallway. The new north facing rear rooms are organised under two folding roof pavilions either side of the circulation space and stair. The concept continues to upstairs creating a central stair and gallery with two bedroom pavilions either side.

The building materials used have been selected to improve the home's sustainability and reduce toxic out-gassing. Examples include NSW grown spotted gum timber and veneers, window frames, decking, shingle cladding.

Name of Project / 项目名称:
Mosman Green
Location / 地点:
NSW, Australia
Area / 占地面积:
353 m²
Completion Date / 竣工时间:
2010
Architecture / 建筑设计:
Anderson Architecture
Landscape / 景观设计:
Jane Coleman Landscape Architect
Interior Design / 室内设计:
Mackenzie Design Studio
Photography / 摄影:
Anthea Williamson Photography

本案是对一个位于莫斯曼的双层小楼项目进行翻新、改建的工程，这是一所属于维多利亚时代晚期或者早期联盟时代的建筑。早先的设计使后面的房间只能依赖其他房间来实现采光和通风。一楼卧室的设计和建造也十分拙劣。

在概念上，现任设计师将新的生活空间和后院连接到一起，从而保留了原来围绕中央入口和门厅的房间。后面朝北的新房间被安排在通风区域和楼梯两边折叠的屋顶下。该理念也被延伸到楼上空间的规划中，在中央楼梯和走廊的两侧分别设立了两间卧室。

建筑材料的选用要提升房屋的可持续性和避免有毒气体挥发。例如，新南威尔士州生长的斑点桉木材、表面饰板、窗框、户外地板和木瓦面等材料都符合这一要求。

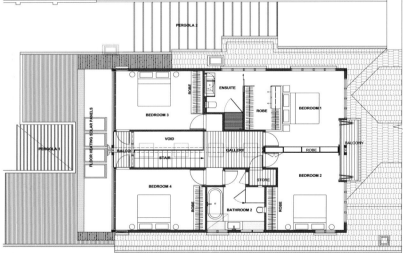

The stair treads and risers as well as floor boards to the first floor are made from recycled timber. New generation water based protective coatings for exposed steel and tung oil based timber finishes reduce the homes environmental impact. Reliance on the town water supply has been reduced with the inclusion of over 36,000L of rainwater storage capacity in tanks located under the rear rooms which supply the washing machine, toilets and garden.

The rear living spaces face north allowing passive solar heat gain in winter through the use of a thermally massive floor structure. When outside temperatures are undesirable, the building can be closed and the large amount of insulated thermal mass helps in keeping the internal temperatures stable. The rear addition is built on a "green star 3" concrete floor structure. The concrete is made from recycled aggregate, fly ash and cement replacement compounds to save 1 tonne of CO_2 per cubic metre compared with standard concrete.

To further warm the new spaces, solar collectors on the roof transfer heated water to the concrete slab below to provide a healthy, radiant heat source. The hydronic system also heats the homes hot water supply with a gas boosted instantaneous gas hot water heater topping up the system when required. The roof is also accommodates a 2.1 kilowatt photovoltaic system to reduce the homes electrical use, and the panels are mounted between the bedroom pavilion gables to reduce their

visual effect. The solar water and space heating systems and the excellent natural ventilation greatly reduce air-conditioning requirements. To further reduce power consumption, the design incorporates extensive strip LED, T5 and compact fluorescent lighting and low power consumption white goods.

一楼的整个楼梯、地板的板材都来自可循环的回收木材。新一代水性工业防腐涂料保护裸露在空气中的钢铁。以桐油为原料的木质漆减少了木材对家居环境的影响。房屋后面贮存雨水的装置可以为这里节约36 000升的生活用水，这些水可以用来清洗机器、卫生间和浇灌花园。

后面朝北的生活区可以在冬天利用被动式太阳能来为地热装置供给能量。如果外面的气温过低，可以将这个区域关闭，这样大量的绝缘蓄热体也可以帮助室内保持一定温度。后面区域的设施是基于"绿色三星级"混凝土地面结构而建。混凝土由可循环的骨料、粉煤灰和水泥制成，较往常的化合物相比，每立方米可以节省1吨二氧化碳。

为了使这个新空间更加温暖，房顶的太阳能集热器将热水转换到下面的混凝土板中，提供了健康的地热辐射热能。液体循环加热系统同样提供热能，当有需要的时候，煤气加热可随时补充进来。屋顶设有2.1千瓦的光电系统，以补充住宅用电，卧室之间安装的镶板减少了可视效果。太阳能热水器、空间供热系统和良好的自然通风都有效地减少了空调装置的使用量。为了进一步减少能量消耗，设计中还使用了大量的LED、T5和密集的荧光灯照明设备以及低能耗的家电产品。

Hidden House Residence

"Hidden House is the perfect example of how incorporating an existing structure can actually be the key to a successful new design", explains Jeffrey Allsbrook, Standard Principal and Co-Founder.

Hidden House is located on a serene site. (In Glassell Park where the paved road ends at an old hand written sign marking the entrance to "Hidden Valley") The property, which can only be accessed via a 800-metre unpaved road, offers expansive views of the city but seems a world away from Los Angeles at the same time.

Anticipating city restrictions associated with building on a site far removed from the street, Standard opted to keep the structure of the existing two-bedroom house substantially intact. At the same time, the architects designed an entirely new home around the original space.

Standard建筑事物所负责人兼创办人Jeffrey Allsbrook说："Hidden House这个项目是诠释如何将原有结构转化为新理念设计的最佳典范。"

Hidden House位于一处宁静祥和的地方（通往格拉塞尔公园的路面尽头有一个很古老的手写标志，这里就是Hidden Valley的入口）走到Hidden House只需800米的路程。从住宅向远处望去，可以将整个城市尽收眼底，同时它又像一个远离旧金山的世外桃源。

由于城市建筑的限制，该住宅被建造在远离街道的地方，Standard团队选择保持现存两个卧室的大体完整。同时，建筑师在原来的空间基础上设计出了一个全新感觉的家。

Name of Project / 项目名称:
Hidden House
Location / 地点:
California, USA
Area / 占地面积:
325.16 m²
Completion Date / 竣工时间:
2009
Architecture / 建筑设计:
Standard Architects
Landscape / 景观设计:
Standard Architects
Interior Design / 室内设计:
Standard Architects
Photography / 摄影:
Benny Chan
Client / 客户:
Cannot Disclose

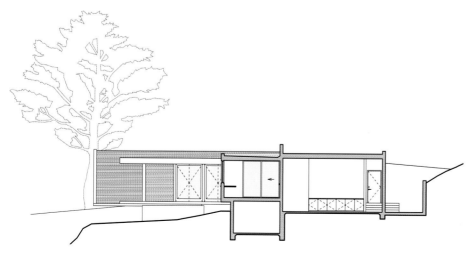

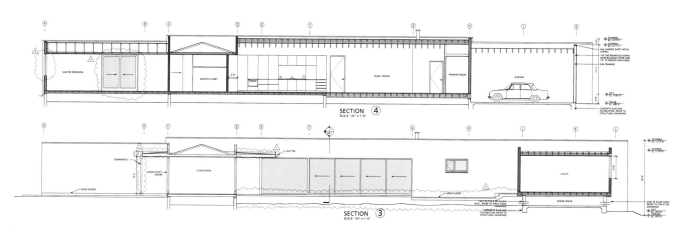

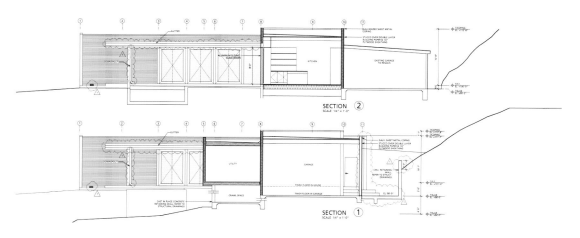

175

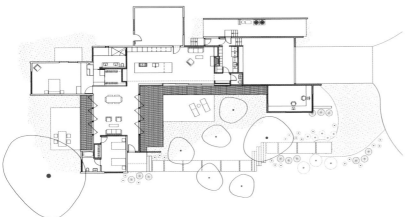

Today, the original two-bedroom cottage is incorporated into the house as the living and dining room. Standard added a new kitchen, family room, office, garage, master bedroom suite and kids bedroom, essentially doubling the volume of the house from 147 to 325 square metres. The new house is arranged around two main courtyards.

The main living spaces open up onto the interior courtyard, while the exterior courtyard looks out over the city in the distance. The self-contained cubes are arranged around the original footprint in such a way that they make order of the disorder. At a later date, the family may add on additional bedrooms per their original plan.

Hidden House also features several sustainable materials and features, ranging from the redwood cladding, to the reclaimed endgrain block wood, to the cork flooring in the office, and to the highly efficient appliances/equipment. The design allows for excellent cross ventilation and day lighting, reducing the need to run forced-air conditioning or heat or energy-consuming electrical lighting. The house is insulated with sustainable cotton and built to be solar-ready. The garden is planted with native landscape and vegetables. Hidden House offers the ultimate country living experience in the middle of an urban environment."

如今，房中原有的两间卧房被改建成起居室和餐厅，Standard团队还为这里添加了新的厨房、家庭娱乐室、办公室、车库、主卧室和婴儿房，从根本上将住宅的面积从147平方米扩大到325平方米，体积增加了一倍多。新建的房屋被安设于两个主要庭院周围。

主要居住空间被安排为开放到内部庭院，同时，从外部庭院可以俯瞰远处的城市。按照这种方式，他们基于原本的布局路线将这个自包含立方体构造重新规划，实现了无序中的有序。不久以后，这个家庭也许会根据他们原来的计划，在这里增加一间额外的卧房。

Hidden House还应用了一些可持续性的材料，无论是红木覆面层、端面回收木块、办公室的软木地板，还是高效能的设施设备，都体现出了可持续性这一特点。良好的对流通风和日光设计大大减少了空调设备、供热和耗能电灯的使用。房内还使用了可持续性的棉花来做绝缘，并充分利用太阳能。花园则种满了本地生长的植物和蔬菜。Hidden House为习惯于城市环境的人们提供了一个极致的乡村生活体验。

Ormond Esplanade House

Judd Lysenko Marshall Architects have designed Ormond Esplanade House in the suburbs of Melbourne, Australia. Make it light. Make it bright. Make it tight. Add all of the green stuff too, please. And, the architects want it beachy. Maybe like a boat? And it better be fun. Serious fun.

This is a sensory architecture that cannot be engaged in a single eyeful. It's a home of full of wonder and anticipation for a young family. A tight wrapped skin creates a shell for this vertical house. Like real skin, it's blemished and rough. Something to protect the delicate inside, at once tough and resilient, but still stretchy, smooth, supple and tactile.

Though seemingly abstract, this house is multilayered, replete with veiled secrets, like a dreaming or wandering left for the occupant to decode – a chain of intimate revelations. A tight block turns the circulation on its head. A three-level stair rises behind the circular windows providing a connection between the four distinct zones. The screen printed hoop pine balustrade links the levels through a carefully constructed narrative of image and colour.

Name of Project / 项目名称:
Ormond Esplanade
Location / 地点:
Melbourne, Australia
Area / 占地面积:
240 m²
Completion Date / 竣工时间:
2010
Architecture / 建筑设计:
Judd Lysenko Marshall Architects
Interior Design / 室内设计:
Judd Lysenko Marshall Architects
Photography / 摄影:
Shannon McGrath

Ormond Esplanade住宅位于澳大利亚墨尔本郊外，由Judd Lysenko Marshall建筑事务所担纲设计。在设计之初，客户提出了这样的要求："轻盈、明亮、紧凑、绿色，同时还要充满海滨风情"。于是，建筑师决定将其建造成一个像小船一样的房子，既环保，又有迥然不同的趣味。

Ormond Esplanade住宅是一座会感知的建筑，人们永远无法用简单一瞥去洞悉它。同时，它还为年轻的房主带来无限的惊喜和希望。建筑外墙被废旧木条紧紧包裹，仿佛是真的皮肤，有效地保护了建筑内部精致的内部空间。

虽然看起来略显抽象，但是整个建筑充满层次感，仿佛是一个暗藏了许多秘密、充满梦幻和惊喜的地方，等待着它的主人去发现。楼梯被设在环形窗户的背后，从一楼一直延伸到三楼，将住宅内四个功能区连接在一起。楼梯的栏杆由南洋杉制成，上面的丝网印刷，无论是构图还是颜色的搭配都十分精细，室内空间愈发和谐。

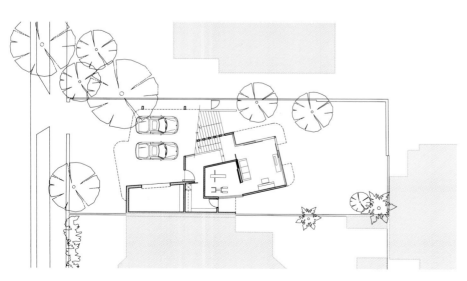

A garden room lands the ground floor. A grand stair is mirrored inside and out and connects the lofty living to both the beach and the backyard. The kid's bedrooms are fun, exciting and colourful and the parent's garret is necessarily moody, sexy and private.

The main stairwell has open risers and ascends through a void which provides passive stack ventilation, grounded in a masonry level and a permeable lid. Glazing is optimised and the daylight is carefully controlled. A 2.0 KW solar system is installed to provide energy. Rainwater flushes and washes while grey water keeps the grass green.

一楼空间设有花园房，其前后墙壁均安装着巨大的玻璃，在玻璃墙的反射下，海边风景和后院的迷人景致尽收眼底。儿童房的设计趣味十足，色彩缤纷。家长的卧室被设在阁楼处，室内设计更加有情调、性感、私密。

楼梯中设有通风管，气流可以通过室内一处大的中空带一路向上，达到被动式通风效果。同时，建筑师优化了住宅内的玻璃装配，方便控制日光直射。住宅安装了一个功率为2.0千瓦的太阳能系统，提供能量。收集而来的雨水可以用做冲洗用水，余下的废水仍可重新利用，用来灌溉草坪。

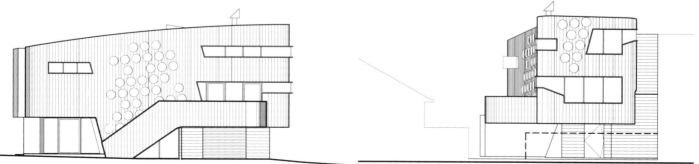

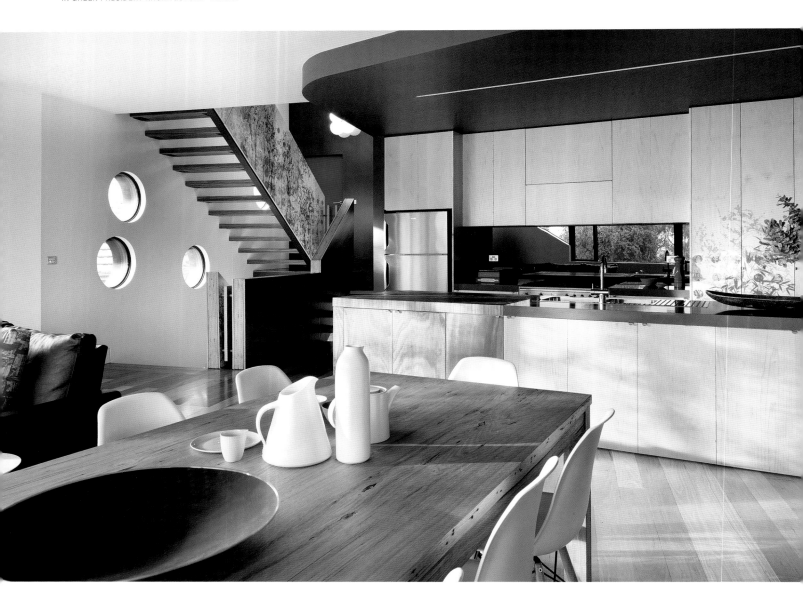

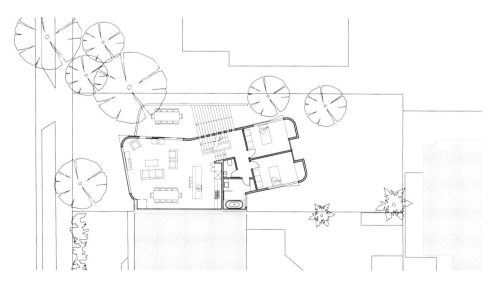

Panel House in USA

The design intent is to create a series of angled walls and reveals in the side elevations in order to provide for view corridors down the side yards to the ocean. The space between the tapered walls is used for pivot windows, which allows for the modulation of the natural prevailing breezes through the house.

Two issues arise from having such a transparent west facing façade, the need for increased privacy, and implementation of solar modulators. A system of aluminum louvers was designed to combat both issues, minimizing the solar gain and providing the desired privacy.

In addition to stairways, vertical circulation is addressed through the use of a glass pneumatic elevator. The elevator is the quickest and easiest access to the rooftop where there are photovoltaic panels, solar panels and a never-ending pool. Space is tight on this narrow lot, and requires the use of every available surface to achieve the sustainability, functionality and the desired quality of life.

本案旨在建筑立面上打造一系列呈角度墙体和外墙与门窗之间的门／窗侧，以便于形成观景走廊，欣赏海滨美景。锥形墙之间位置被用于安装中悬窗，这样的设计可以调节穿过空间的自然风量。

住宅西侧的透明设计引起两个问题：其一是如何提高住宅的私密度；其二是如何安装太阳能调制器。于是，建筑师设计了一个铝制百叶窗系统，令以上两个问题迎刃而解。该百叶窗系统不仅可以将室内光照最小化，还可令空间更加私密。

除了楼梯外，Panel House还拥有一个玻璃气动电梯，人们可以用最快的速度和以最便捷的方式通向屋顶。屋顶安装着光电板、太阳能板和一个水池。因为空间相对紧凑，建筑师必须充分利用每一处可以利用的表面，以得到一个集可持续性和功能性为一体，同时可以为业主提供理想生活质量的住宅空间。

Name of Project / 项目名称:
Panel House
Location / 地点:
Venice, California, USA
Completion Date / 竣工时间:
2006
Architecture / 建筑设计:
David Hertz Studio EA
Landscape / 景观设计:
Thomas Ennis
Interior Design / 室内设计:
Thomas Ennis
Photography / 摄影:
Juergen Nogai
Client / 客户:
Thomas Ennis

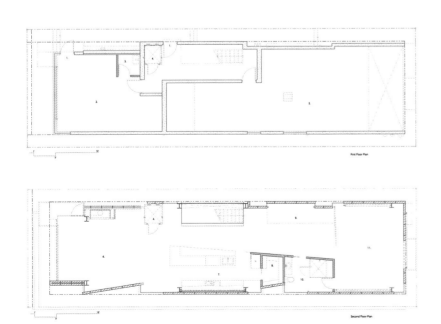

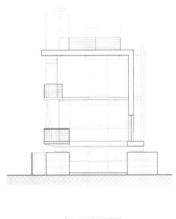

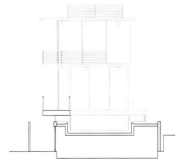

The solar system comprised of 14 south-facing Photovoltaic panels and an inverter, production 2.3 kWh of energy per day – sending energy back to the grid. Solar collector provides hot water. Louvers are used in shade applications as blinds and window coverings to direct views as well as minimizing solar gain.

Automated skylights for natural ventilation control – programmed on set point thermostatic and humidistat control. Pivot windows are controlled manually to modulate airflow from the ocean – opening up the house to the prevailing breeze for natural ventilation. Staircase acts as a solar chimney, and the hot air rises through the open space and exits at the top through the skylight.

High performance pre-fabricated manufactured refrigeration panels are used to create the exterior walls. The panels are coated with aluminum sheets, which is the final finish for both the interior and exterior. There is no wood framing in the house. FCS certified walnut cabinetry and counters are used in the kitchen and bathroom.

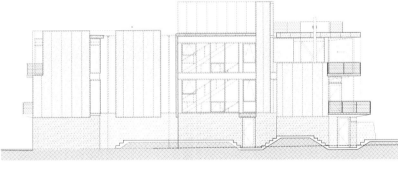
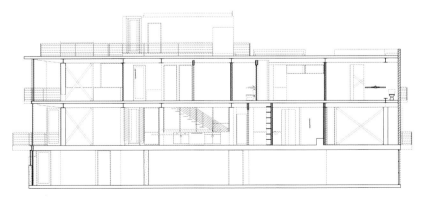

太阳能发电系统由14片朝南放置的光电板和一个逆变器组成，该系统每天可以产生2.3千瓦时的电能，并将剩余能量重新发回到电网中。太阳能集热器解决了住宅的热水供给问题。百叶窗的使用不仅可以保护室内隐私，同时可减少光照度。

自动天窗用来控制自然通风，进而调节室内温度和湿度。房内还装有可以手动控制的中悬窗，用以调节来自大海的气流，令空间获得更为自然的通风。室内楼梯间可以起到烟囱效应，引起空气对流，令热气上升并通过天窗排出。

预制高性能制冷板被用于创建建筑外墙，这些制冷板表面均镀有铝片，作为室内、外空间的最终饰面。出于防火性能的考虑，住宅内未采用任何木质框架。厨房和浴室内所用的胡桃木细木家具和柜台全部经过FCS认证。

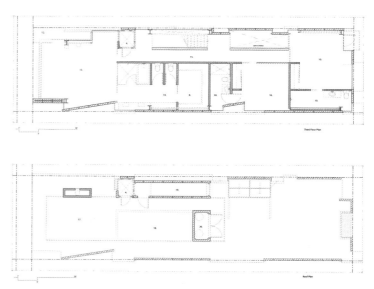

193

Rozelle Green in Australia

This project involved alterations and additions to a north-south-oriented terrace on a double width block. Renovations in the 1980's vertically separated the rear south-facing living spaces from the yard, with service spaces directly adjacent to the yard, allowing little internal-external connection and a cold and dark internal environment.

The new design lowers the living spaces to better connect with the landscape and incorporates sustainable principles to reduce the home's ecological footprint. The new family room adjacent to the rear yard utilises sustainable spotted gum hardwood window frames and lowe-glazed doors that open on two sides, making the room part of the landscape when open.

When outside temperatures are undesirable, the building can be closed and the large amount of insulated thermal mass keeps the internal temperatures stable. The rear addition is built on a "green" concrete floor structure coupled to the ground.

Name of Project / 项目名称:
Rozelle Green
Location / 地点:
NSW, Australia
Area / 占地面积:
222 m²
Completion Date / 竣工时间:
2009
Architecture / 建筑设计:
Anderson Architecture
Landscape / 景观设计:
Eximia Design
Interior Design / 室内设计:
Anderson Architecture
Photography / 摄影:
Nick Bowers Photographer,
Steve Brown Photography

本案是对一个南北朝向的住宅进行翻新与扩建。在20世纪80年代的一次改建中，建筑师在垂直方向上把位于住宅后侧的朝南生活空间同院子分离出来，把综合服务空间安排在院子旁边。这样的设计将室内空间和室外隔离开来，令室内空间变得有些阴冷。

在本次翻新设计中，来自Anderson建筑事务所的设计团队降低了生活空间所在的位置，使其更好地与所在景观联系在一起。这样的设计也利于可持续性策略的融入，减少对环境的影响。新的家庭居室毗邻后院而建，使用了可持续性的斑点桉硬木窗框和低辐射玻璃门。空间两侧十分通透，当玻璃门完全打开时，家庭居室便成了周围景观的一部分。

当外部温度过冷或过热时，可以将门关起来，建筑中大量使用的热绝缘体可以使室内温度维持稳定。住宅后侧的新增结构被建于"绿色"的混凝土地面结构上，给予地面双重保护。

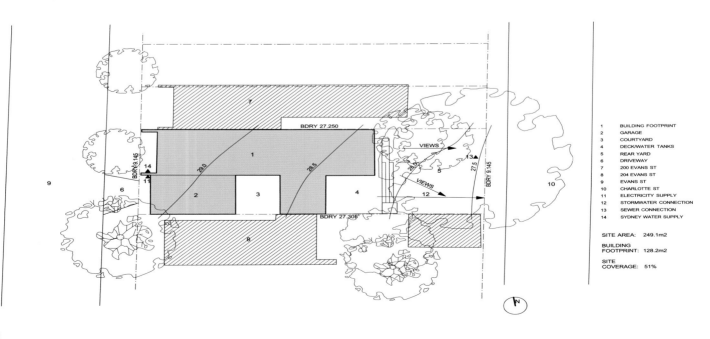

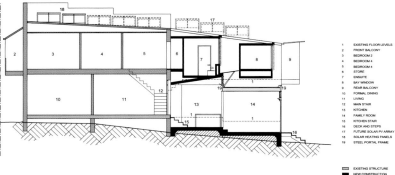

1	EXISTING FLOOR LEVELS
2	FRONT BALCONY
3	BEDROOM 2
4	BEDROOM 4
5	BEDROOM 4
6	STORE
7	ENSUITE
8	BAY WINDOW
9	REAR BALCONY
10	FORMAL DINING
11	LIVING
12	MAIN STAIR
13	KITCHEN
14	FAMILY ROOM
15	KITCHEN STAIR
16	DECK AND STEPS
17	FUTURE SOLAR PV ARRAY
18	SOLAR HEATING PANELS
19	STEEL PORTAL FRAME

EXISTING STRUCTURE
NEW CONSTRUCTION

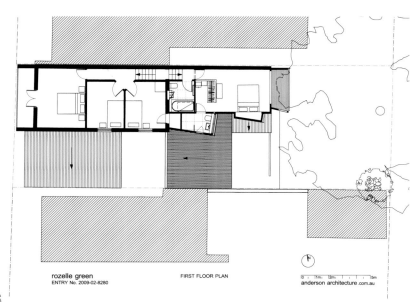

rozelle green
ENTRY No. 2009-02-8280

FIRST FLOOR PLAN

anderson architecture.com.au

The building materials used have been selected to improve the home's sustainability and reduce toxic out-gassing. Examples include eco certified timber veneers for kitchen joinery, sustainably managed hardwood timber windows, decking, deck framing and weatherboards, LOSP treated plantation pine framing, skirtings and reveals, low VOC timber finishes and new generation water based protective coatings for exposed steel.

Day time living areas located on the ground floor have excellent ventilation through bi-fold doors and highlight louvres as well as convective ventilation though 1st floor windows when the interconnecting doors are opened. Winter heating is supplied by solar powered hydronic floor heating in this area and the bedrooms above can benefit from this source at night as the thermal mass releases energy and the heat rises.

Due to the south-easterly aspect of the house, the need for an air-conditioning system has been eliminated through the incorporation of adequate thermal mass, window shading, high levels of the insulation and the very efficient natural ventilation system.

Winter solar exposure has been maximised through the use of well proportioned eaves and retractable external louvres. An over-sized array of solar collectors supplies heating to the concrete floor slab and maximum benefit of this heating is gained through the use of high insulation levels and natural ventilation to further circulate the warm air to other zones.

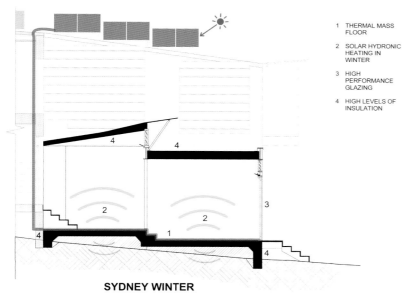

SYDNEY WINTER

1. THERMAL MASS FLOOR
2. SOLAR HYDRONIC HEATING IN WINTER
3. HIGH PERFORMANCE GLAZING
4. HIGH LEVELS OF INSULATION

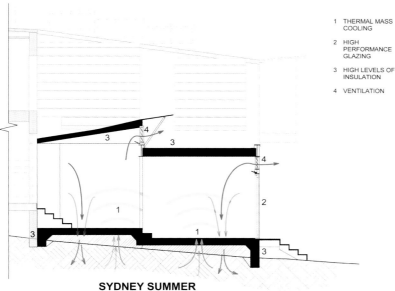

SYDNEY SUMMER

1. THERMAL MASS COOLING
2. HIGH PERFORMANCE GLAZING
3. HIGH LEVELS OF INSULATION
4. VENTILATION

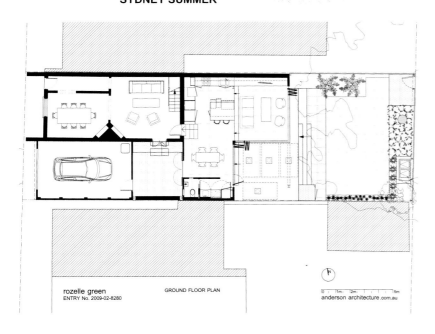

rozelle green
ENTRY No. 2009-02-8280
GROUND FLOOR PLAN
anderson architecture.com.au

为提高住宅的可持续性，降低有毒性气体的挥发，本案所用材料都是经过精心挑选的，例如，厨房中细木家具所用的经过生态认证的木材贴面和家庭居室中经可持续处理的硬木窗户、户外地板、边框以及挡风板等。还有经有机溶剂防腐剂(LOSP)处理过的人工种植松木框架、壁脚板、窗侧，低VOC（挥发性有机化合物）木质饰面及外露钢材所用的新一代水性防护涂料。

日间生活区位于一楼，通过双折门和突出的百叶窗，实现了良好的通风，此外，当互相连接的门被打开时，通过楼上落地窗，亦可实现对流通风。该区域的冬季采暖由太阳能液体循环加热地板提供，到了夜晚，蓄热体会持续释放能量，加之热气上升的原理，位于客厅上部的卧房仍可获取热量供给。

由于建筑朝东南方向，因此只需采用足量的蓄热体、遮窗体、高隔温材料和高效通风系统，就可以免去空调系统的使用。

比例恰到好处的屋檐和伸缩自如的外部百叶窗，令住宅在冬季时所接收的阳光照射面达到最大。一排超大型的太阳能集热器阵列将热能传输到混凝土楼板中。住宅较高的绝缘性和良好的自然通风，非常便于温气流进一步传送到房间其他区域，实现了热能利用率的最大化。

Setia Eco Villa in Malaysia

This project of 400 square metres is located at the palm coconut plantation, in Syah Alam district, 15 minutes ride from Kuala Lumpur. The site where the building located is part of the master plan of satellite city, which comprises of residential, and other supported facilities, named Setia Eco Park.

This development bring the Green Environment Concept, within the site planning, dedicated to middle-upper class market. The client's brief is to create a unique, modern, sustainable architecture style without leaving tropical aspects as a starting point.

The usage of few kind of texture and colour to this building, also refers to the richness and diversity of Asian architecture in general.

本案由TWS & Partners团队担纲设计，占地400平方米。位于Syah Alam地区的棕榈椰子种植园内，距离吉隆坡仅15分钟车程。该地段是卫星城总体规划的一部分，由住宅区及配套设施Setia生态公园共同组成。

在场地规划中该项工程融入绿色环境概念，主要面向中上层消费群体。客户意在打造独特、时尚、可持续性建筑，构建展现亚洲建筑风格的时尚家居环境，而热带因素不会成为设计的局限。

本案为数不多的色彩及材料的运用展现了亚洲建筑在总体风貌上的丰富多样性。

Name of Project / 项目名称:
Setia Eco Villa
Location / 地点:
Kuala Lumpur, Malaysia
Area / 占地面积:
400 m²
Completion Date / 竣工时间:
2007
Architecture / 建筑设计:
TWS & Partners
Photography / 摄影:
Courtesy of SP Setia
Client / 客户:
SP Setia

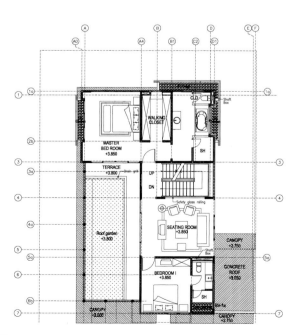
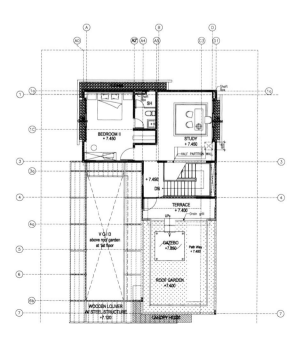

The passive instrument is used to create a sustainable architectural space. Derive from the client's brief, architects decide to play an additive form generator to create a massing and, at the same time, functional space as well. The indoor space is analogized in pure geometric box, which composed, juxtaposed and inter connected one and the other with a "gap" in between, which interpreted to a circulation or buffer zone, to infiltrate natural air and daylight, as much as possible.

In vertical way, the massing are rotated and juxtaposed perpendicular one and the other to create the in-between space into roof garden. With this roof garden, architects try to create a rain water catchments surfaces to replace the ground which covered by the building. And give an added value to the room at upper level, as a direct outdoor orientation from within. This rotation of the massing also gives a shadow effect to the room below.

The usage of wide canopy in the top part of the building also plays a double role, as a place for solar panel, beside as a canopy. This amount of solar panel is calculated to reduce 30% of energy consumed within this villa.

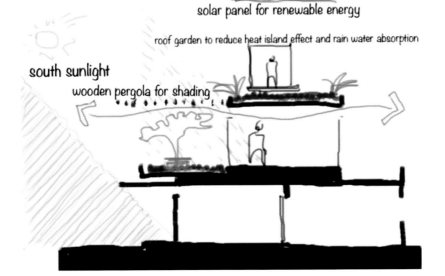

建筑师以被动式节能举措为业主，打造出可持续性建筑空间。基于业主的要求，建筑师决定通过附加形式设计，营造功能性空间。他们将室内空间打造为纯几何方块构造，各区域互相并列又通过之间的"缺口"紧密相连。以此形成流通区或缓冲区，最大限度地吸收自然光线，保证室内空气流通。

在垂直方向，建筑体块旋转呈并列垂直分布，中间区域则被打造为屋顶花园。有了这个屋顶花园，建筑师尝试创建一个雨水集水区，以取代被建筑物覆盖的地下装置，它作为户外一处景观，为楼上房间带来一定附加值。这种旋转布局也给楼下房间带来遮阴效果。

建筑顶部的宽阔檐篷具有两方面作用，在作为檐篷之外，也用做放置太阳能电池板的场所。这里应用的太阳能电池板的数量将为别墅节省30％的能源消耗。

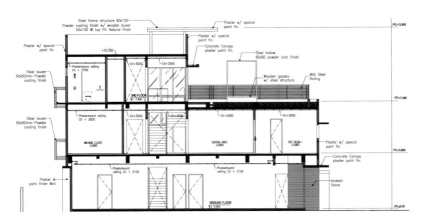
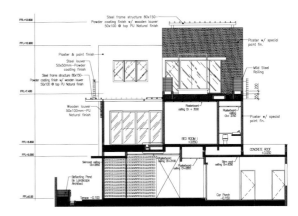

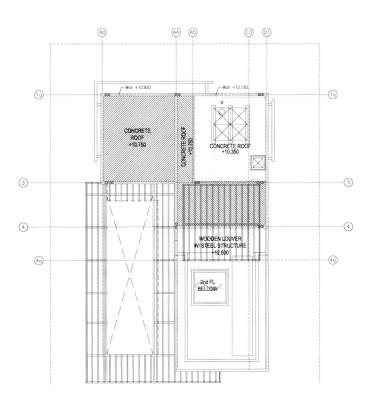

Sideris House in Greece

This three-level house is located in a suburb of Larissa, Greece. The building consists of several different volumes witch play with space. The configuration of multi-flight staircases which combine series of steps and rest areas, predominate the facade of the building as well as the inner space.

The high ceilings help the staircase to expand its elegance to a maximum. A well-planed lighting scheme emphasizes the staircase's form and the quality of the materials. Metal construction can attain and it endow the staircase with sculptural qualities giving it a starring role in the building over and above their practical function. The ground floor houses an artistic studio, a small inside garden and the entrance stair case which leads to the first floor. There the living room, the dining room, the kitchen area and a small bathroom are harmonically situated. On the upper floor, people will come "face to face" with the glass corridor which surrounds the building and leads to the master bedroom, the laundry room and the master bathroom.

The two glass facades give a sense of closeness to the green exterior and combined with the glass balcony, the flying-like staircases and the high ceilings create the perfect atmosphere for a relaxed and thought-promoting atmosphere.

该三层家居建筑位于希腊拉里萨郊区。房内设有功能各不相同的若干区域。多梯段楼梯的设计将一系列台阶、休息区、占主导地位的建筑立面以及室内空间贯穿为一体。

高高的天花板令楼梯充分展示其高雅的魅力。精心设计的照明设备突出楼梯的结构及建筑材料。金属结构赋予楼梯一种雕刻般的特质,在实用功能基础上成为整个建筑中的亮点。一楼设有充满艺术气息的工作室、小型室内花园,以及在入口处通向二楼的楼梯。楼梯通向楼上的客厅、餐厅、厨房及一间小浴室。再往楼上,则会看到"面对面"式的玻璃走廊将建筑环绕其中,通向主卧、洗衣房及主卫。

两面玻璃幕墙与玻璃阳台完美融合,营造出与室外绿色景观的紧密相连之感。颇具飞翔之感的楼梯与高高的天花板营造出完美的舒缓氛围,给人以无限遐想。

Name of Project / 项目名称:
Sideris House
Location / 地点:
Larissa, Greece
Area / 占地面积:
285 m²
Completion Date / 竣工时间:
2008
Architecture / 建筑设计:
Christina Zerva Architects
Interior Design / 室内设计:
Christina Zerva Architects
Photography / 摄影:
Mihajlo Savic
Client / 客户:
Mr.Sideris

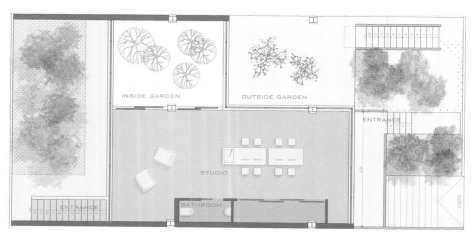

For construction, architects chose steel for its sustainability, refurbishment, recyclability and reusability. Moreover, the seismic behavior of steel is far more satisfactory than that of concrete. They opt for Ytong blocks since they are fire resistant and convenient for thermal insulation. It is ecological and durable resulting in lower CO_2 emissions.

Sunlight is converted into electricity by photovoltaic solar panels hidden in the roof top. Solar thermal energy, heating water with energy from the sun, is an entirely renewable, reliable and cost-efficient energy source, that can be replenished easily.

Recycled and second hand materials have been used for the building. Reuse of old building materials save energy that is spent in manufacturing and transporting. The building's wide windows and height ceilings allow natural available forces to supply and remove air through the enclosed space, improving or maintaining the quality of the air providing acceptable indoor air quality.

建筑师选用钢铁作为建筑材料，在可持续、可翻新、可循环及可再利用等方面性能优越，而且钢铁的抗震性能远比混凝土好。建筑师选用抗燃、绝缘性能比较好的YTONG轻砂加气混凝土块为材料，建成了既生态环保、又可持久利用的低碳建筑。

屋顶的光伏太阳能板可将日光转变为电能。源自日光的能量也可将水加热，是一项完全可再生、可信赖、且具有经济性的能源，并方便得到补充。

回收和二手材料已用于本案建设中，旧建材再利用的节能性体现在生产和运输方面。宽敞的窗户及高高的天花板允许房屋利用自然力量实现良好的空气流通，有效保证室内空气质量。

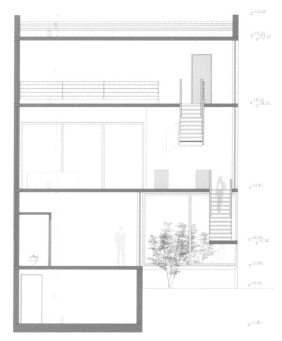

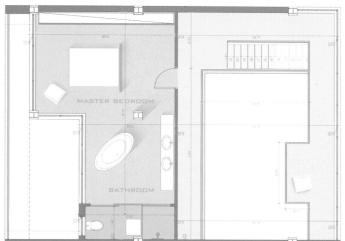
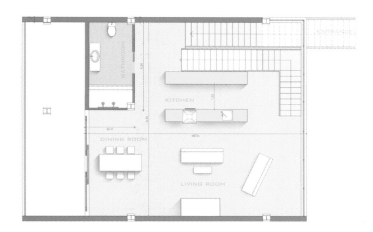

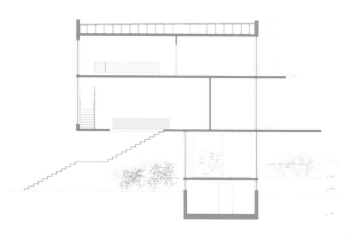

The House The Slope in Russia

Totan Kuzaembaev Architectural Workshop completed The House The Slope in 2007. With an area of 472 square metres, the project is located in the private resort of Pirogovo in Moscow region, Russia. Totan Kuzaembaev Architectural Workshop uses not only sustainable materials as building materials, but employs passive energy measures for the project. The unique building shape highlights the project and reminds people of sea animal, devilfish.

The unique atmosphere of the resort develops from the nature as unspoiled as possible and the acutely modern nature-consistent architecture freely placed in a natural environment. The wooden Devilfish House is located on a glade in beautiful relic pine wood on the bank of a gulf. The shape of the meadow intended for building reminded the architect of some sea animal, a ray or probably devilfish, and gave him idea of the shape of the plan.

Located in a natural environment, The House The Slope merges with the surrounding atmosphere naturally. In addition, one more remarkable line: there is no fence around this private house, as well as around the other houses in the resort territory, which is absolutely unusual for Russia.

The House The Slope由Totan Kuzaembaev Architectural Workshop担纲设计，于2007年竣工。本案位于俄罗斯莫斯科地区的Pirogovo私人度假区内，占地472平方米。Totan Kuzaembaev Architectural Workshop使用可持续性材料作为本案的建筑材料，并采用被动节能措施。独特的建筑外形是本案设计的亮点所在，令人们不禁联想到海洋动物章鱼的形状。

度假区的独特氛围源自尽可能未受污染的自然地理环境，堪称是一座现代与自然完美融合的高尖端建筑之城。这座章鱼木房建于海岸旁美丽的松木遗迹间一处林间空地上。该建筑选址的空地形状令建筑师联想到海洋动物鳐鱼或章鱼的形态，其设计灵感正是得益于此。

The House The Slope位于自然环境之中，与周围氛围完美融合。此外，更加醒目的一处亮点在于这座私人住宅周围没有围栏，Pirogovo度假区内的其他住宅也没有任何围栏，这在俄罗斯地区是极为罕见的。

Name of Project / 项目名称:
The House The Slope
Location / 地点:
Moscow, Russia
Area / 占地面积:
472 m²
Completion Date / 竣工时间:
2007
Architecture / 建筑设计:
Totan Kuzaembaev Architectural Workshop,
Kuzembaev Totan,
Savanec Sergei,
Salina Mariya
Landscape / 景观设计:
Totan Kuzaembaev Architectural Workshop
Interior Design / 室内设计:
Totan Kuzaembaev Architectural Workshop
Photography / 摄影:
Ilya Ivanov,
Courtesy of Building ARX

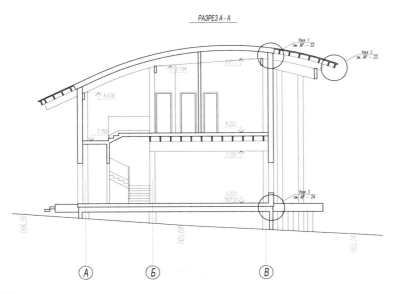

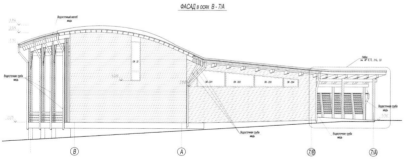

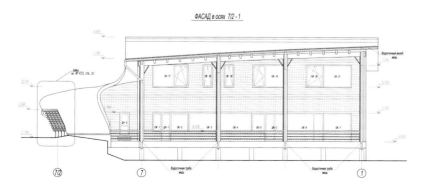

The main intention during the design process was to avoid any damage to the trees (no trees were destroyed during the construction). A frame made of glued wood racks (made from spruce which is a local construction material) is a constructive base of the building. The finishing material adopted is larch. Different functions are combined in this two-storey house. Residential space and the garage are situated under one roof. This roof is going from one volume to another following the steps of the floor levels of different parts of the building. The main part of the building is installed on the piles. Such approach has allowed leaving untouched a little elevation of relief underneath the building.

The huge two-level open terrace on the main facade is another expressive element uniting the house with the nature – some kind of intermediate, transitive space. Orientation of this almost fully-glazed facade to the south allows saving on energy for the heating.

The timber – wooden tiles (shingle) – is also used as a roofing material. As a result, the curved roof of the building reminds scales of a strange sea animal who has crept out onto the shore to bask in the sunshine.

建筑师在设计过程中主要避免对树木造成破坏（在实际施工过程中树木没有受到任何破坏）。胶合木板框架（由当地建筑材料云杉木制成）成为建筑的主体构造。所用饰面材料为落叶松木。该二层住宅建筑具有不同的功能区。居住空间与车库被安设于同一层。屋顶形状根据建筑各部分楼层高度而起伏变化。建筑主体部分由桩状物支撑，设计中通过运用这种方法将建筑整体提高，与地面保持一段距离。

在建筑正面一侧巨大的双层开放式露台成为又一具有表现力的元素，作为过渡空间，将住宅与自然完美融合。建筑南面装有许多宽大明亮的玻璃窗，利于节省采暖热能。

屋顶选用木料——木砖（瓦）为材料。弧形屋顶的整体形状令人们联想到偷偷溜到海岸上晒太阳的奇特的海洋动物。

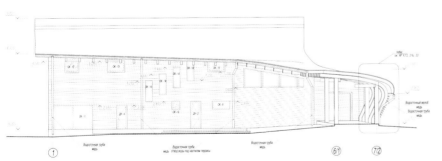

ФАСАД в осях 7/А - В

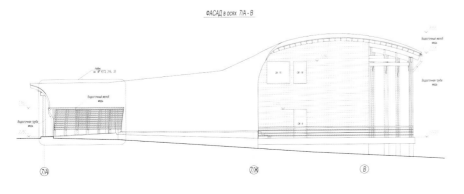

Sorrento Beach House

Marcus O'Reilly Architects has carefully designed a low key, site responsive beach house in Sorrento on Victoria's Mornington Peninsula in Australia. The timber clad structure reinterprets a longstanding beach house tradition in the area. It uses various local materials.

The main living level stretches east to west along the site to welcome northern light into each room of the house. The light is cleverly filtered through tea tree stake sunshades and screens which add to the vernacular of the house. The point of entry of the house is up a timber staircase with playful and unpretentious custom galvanized steel balustrades which leads to an expansive deck covered with a radial polycarbonate and timber awning.

Catching sun and views from a large elevated deck, this beach house has many typical local features: It has sandstone base walling, the weatherboard construction is up on stilts, and it has a skillion roof form. At the same time, stainless steel wire balustrade and a "hole punched" galvanised steel – sided timber stair all combine to give an informal air to this new take on an old theme.

Name of Project / 项目名称:
Sorrento Beach House
Location / 地点:
Victoria, Australia
Completion Date / 竣工时间:
2008
Architecture / 建筑设计:
Marcus O'Reilly Architects
Photography / 摄影:
Dianna Snape

本案位于澳大利亚维多利亚州莫宁顿半岛的Sorrento区，由Marcus O'Reilly建筑事务所精心打造而成，住宅形式低调，和谐地与周围环境结合在一起。木材覆盖的结构重新演绎了当地海边别墅的悠久传统。住宅所用材料均取材于当地。

住宅的主要生活空间顺应其所在的地形，由东向西延伸，方便来自北面的光照直接射入住宅的每一个房间。百叶窗和遮阳屏由当地色彩浓郁的茶树材料制成，令光线巧妙地渗入到空间的各个角落。通向住宅入口的楼梯使用木质材料，扶手由镀锌钢制成，有趣而不失含蓄。经由楼梯向上，是一个宽阔的露天平台，天台上的雨篷由聚碳酸酯和木材制成，业主可以在天台上享受日光，欣赏海边美景。

Sorrento海滨别墅拥有许多当地建筑独有的特色，例如，建筑的底部墙体由砂岩砌成，墙体装有护墙板，整个建筑采用架空设计形式，以及屋顶采用澳洲独有棚屋形式等。同时，不锈钢丝栏杆和打孔镀锌侧梯扶手更为这个以怀旧为主题的建筑带来全新的别样风情。

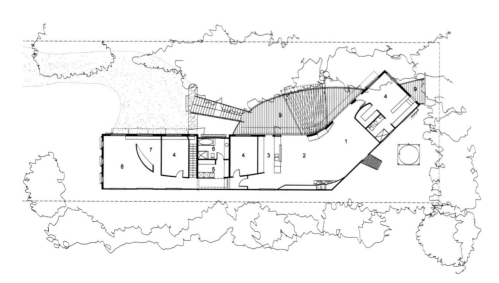

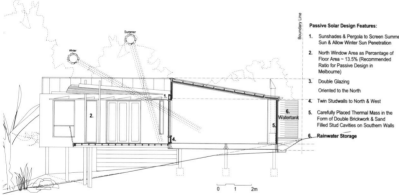

Passive Solar Design Features:
1. Sunshades & Pergola to Screen Summer Sun & Allow Winter Sun Penetration
2. North Window Area as Percentage of Floor Area ~ 13.5% (Recommended Ratio for Passive Design in Melbourne)
3. Double Glazing Oriented to the North
4. Twin Studwalls to North & West
5. Carefully Placed Thermal Mass in the Form of Double Brickwork & Sand Filled Stud Cavities on Southern Walls
6. Rainwater Storage

The architects chose a low maintenance material palette which was not only intended to stand up to the tests of sandy boogie boards, wet towels and teenage parties but also improve with age.

Sustainable issues were to the fore, primarily with passive Solar performance. The orientation, size, volume of building, insulation and the weight were modelled to assess how much extra ballast would be needed to add to the southern walls to obtain the required thermal mass. The calculations paid off and rendered air conditioning unnecessary despite the obvious difficulties involved in obtaining optimal thermal performance for a structure that is up on stilts and lightweight in a temperate climate. Passive ventilation and 10,000 litre rainwater tanks complete the package.

建筑师挑选了一系列维修方便的材料，这些材料不仅要能经得住来自沙滩冲浪板、湿毛巾以及年轻孩子举行的各种派对的重重考验，同时还要经久耐用。

可持续性是建筑师重点考虑的问题。住宅装有被动式太阳能装置，建筑的方向、大小、体积、绝缘性以及重量等因素的设计均为南侧墙体如何获得足够热能服务。尽管对于这样一个拥有架空结构的轻巧建筑，想要在如此温和的气候下，获得最佳的热性能会面临许多困难，但经过严密的计算后，建筑师还是提出住宅可以不安装空调。被动通风设备以及10 000升雨水收集槽的安装，令建筑师提出的设计方案更加可行。

Stony Point House in USA

The design of the house, located outside Charlottesville, Virginia integrates green design strategies on a wooded site on a west-facing hillside. Working closely with the landscape architect, Christopher Hays conceived the house as a series of terraces which were carved into the hillside capped with a roof with a broad overhang. The upper level of the house has a broad prospect of the property.

It is designed for a pair of expatriates, a stained glass artist and an attorney. Having traveled widely, the couple expressed an interest in a quiet contemplative Zen-like lifestyle with a house that allows for the greatest appreciation of their beautiful 12,141 square metres woods.

The house is oriented exactly on the north-south axis. Besides working with passive solar strategies, other green design systems include structural insulated panels for the walls, radiant floors for the lower level heating and solar thermal panels for the hot water.

本案位于弗吉尼亚州夏洛茨维尔以外，建于西向山坡上一处树木繁茂的地方，整合了绿色设计策略。Christopher Hays与景观设计师紧密合作，将房屋构想成嵌于山坡之上的一系列阶梯看台，屋顶则以宽阔的屋檐进行覆盖。房屋的楼上空间具有宽阔的视野。

本案专为一对外籍夫妇设计，他们是彩色玻璃艺术家和律师，有着广泛的旅行阅历。二人对具有禅意的安静生活方式颇感兴趣，因此设计师在这片美丽的12 141平方米美丽的林地中为他们打造了一处安居之所。

房屋恰为南北朝向。除了应用被动式太阳能技术策略，其他的绿色设计系统还包括墙壁上的结构绝缘板、可以为低楼层提供热量的辐射热能地板及用于水加热的太阳能光热板。

Name of Project / 项目名称:
Stony Point House
Location / 地点:
Virginia, USA
Area / 占地面积:
234 m²
Completion Date / 竣工时间:
2009
Architecture / 建筑设计:
Hays + Ewing Design Studio
Landscape / 景观设计:
Nelson Byrd Woltz Landscape Architects
Client / 客户:
Prakash Patel

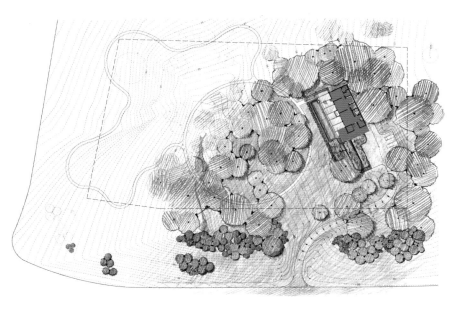

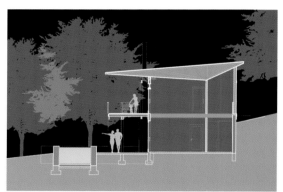
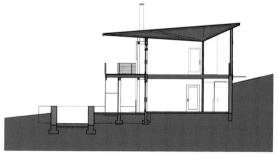
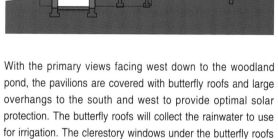

With the primary views facing west down to the woodland pond, the pavilions are covered with butterfly roofs and large overhangs to the south and west to provide optimal solar protection. The butterfly roofs will collect the rainwater to use for irrigation. The clerestory windows under the butterfly roofs will allow for southern and northern light in a house aligned on the north-south axis.

Stained glass by the artist/client is integrated into the clerestories to create a colorful play of light throughout the upper floor. Likewise, the lower level ceiling has a play of afternoon sunlight reflected from the lap pool.

从西部向林地水池方向望去，楼顶采用蝶形屋顶，西南方向巨大的悬壁结构充分利用太阳能资源。蝶形屋顶的设计便于收集雨水，用以灌溉。本案按照南北轴线方向布局，屋顶下的天窗设计利于从南北方向最大限度地吸收自然光线。

天窗上装有房屋主人设计的彩色玻璃，令楼上区域呈现出多彩的光线折射。同样，楼下天花板也会在午后时间折射出由小型泳池反射回来的光线。

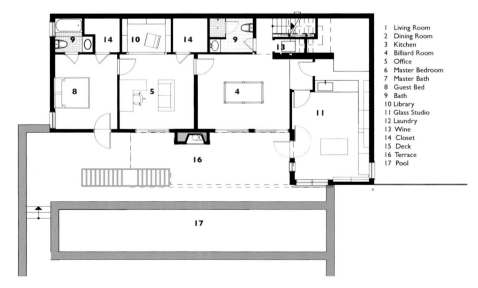

1 Living Room
2 Dining Room
3 Kitchen
4 Billiard Room
5 Office
6 Master Bedroom
7 Master Bath
8 Guest Bed
9 Bath
10 Library
11 Glass Studio
12 Laundry
13 Wine
14 Closet
15 Deck
16 Terrace
17 Pool

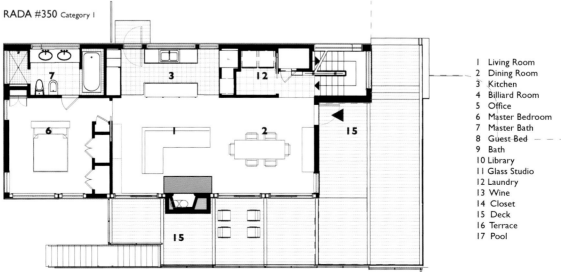

RADA #350 Category 1

1 Living Room
2 Dining Room
3 Kitchen
4 Billiard Room
5 Office
6 Master Bedroom
7 Master Bath
8 Guest Bed
9 Bath
10 Library
11 Glass Studio
12 Laundry
13 Wine
14 Closet
15 Deck
16 Terrace
17 Pool

Telescope House in Russia

The individual house with a wharf is located at water bank. The house is one of the most eccentric constructions of the Pirogovo resort. The motives of Russian architectural avant-garde are cleverly played up here – the diagonal and the cylinder set out on the facade remind the Supremacists' sculptures of 1920s. But, unlike its predecessors, Telescope House is not focused at all on the proletarian asceticism. The expression of its forms reflects the dynamical way of life of the owner-yachtsman.

A reinforced concrete slab resting on metallic piles serves as the foundation for the building. The complicated shape of the building is realized on the base of glued wooden structures covered with larch boards with further refinement. The roof of the building is made of copper. Triple-glazed windows are used. The finishing of the house both inside and outside is performed with the highest quality, close to the quality of the best furniture, or even higher.

这栋带有码头的私人住宅建于水岸边。这栋房子是Pirogovo度假区内最为与众不同的建筑之一。本案充分反映俄罗斯先锋建筑的设计理念——建筑外观上的斜线元素和圆柱体的设计令人们联想到20世纪20年代至上主义风格的雕塑艺术。然而，与之前的建筑不同，Telescope House的设计完全没有体现无产阶级的苦行主义思想。这种建筑形式充分表达了游艇驾驶者充满生气的生活方式。

金属桩上的钢筋混凝土板成为这座建筑物的地基。复杂的建筑构造形状在胶合木板结构结合落叶松饰面的进一步细化装饰中得以实现。建筑屋顶以铜为材料建成，玻璃窗安置三层玻璃。室内选用最高品质的家具，房子室内、外饰面也都以等同、甚至更高质量水准进行施工。

Name of Project / 项目名称:
Telescope House in Russia
Location / 地点:
Moscow, Russia
Area / 占地面积:
128 m²
Completion Date / 竣工时间:
2005
Architecture / 建筑设计:
Totan Kuzaembaev Architectural Workshop,
Kuzembaev Totan,
Kondrashov Dmitry
Landscape / 景观设计:
Totan Kuzaembaev Architectural Workshop
Interior Design / 室内设计:
Totan Kuzaembaev Architectural Workshop
Photography / 摄影:
Ilya Ivanov, Yuri Palmin

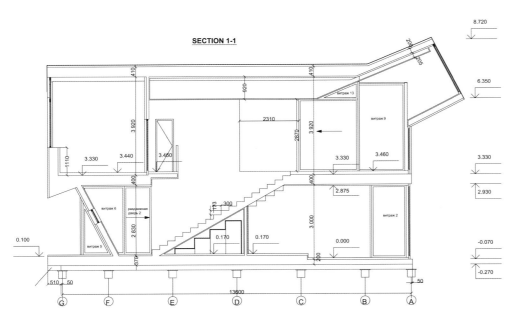

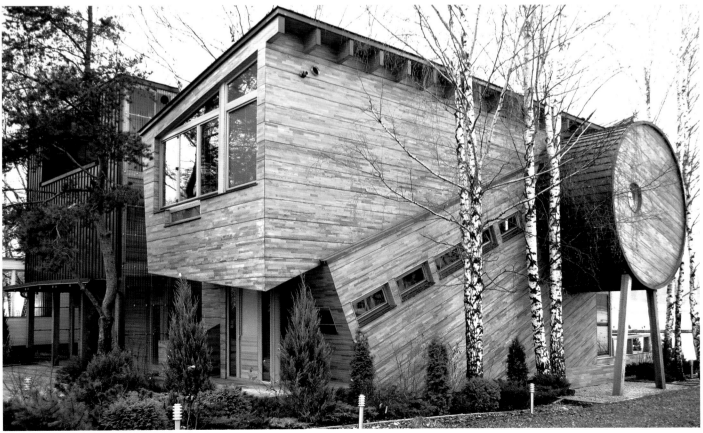

Each element of the house has a specific function. The diagonal turns out to be a ladder with the glazed end face, through which a rising person can see the sky (the name of the house comes from this). And there is an office located in the cylinder. This complicated constructive element with the bent stained-glass window produced with the technique of automobile glasses, gives the exhaustive idea of the high level of the construction.

The architects take the objective conditions into consideration and choose timber as the main building material. They focus on low maintenance and use materials with a low budget. The house is built of glued timber covered with larch laths which has the minimum effect to the environment and also expresses the natural environmental-friendly idea. The inner space is perfectly lit through the big windows facing the bay that provide long hours of natural light and full integration with the surrounding scenery. The terrace extends to the little private quay with a berth.

本案的每一个设计元素都具有一种特定的功能。倾斜元素充当梯子的功能，其端面精致光滑。人们走到上面，可以欣赏到天空的景象（本案名字来源于此）。圆柱体元素的内部设有办公室。这个复杂的构造元素上的弯曲彩绘玻璃窗借鉴于汽车玻璃的制作技术。此处设计充分展现了高水平的建筑理念。

建筑师将客观条件考虑在内，选用木材为本案的主要建筑材料。他们关注维护费用低、预算成本低的材料。因此，胶合落叶松板条成为首选材料，这种材料将对环境产生最小的影响，同时也契合天然环保理念。面朝海湾的巨大玻璃窗使室内空间通透明亮，保证室内得到长时间的自然光照，并使之与周围景色完美融合。露台一直向前延伸到可供泊船的小型私人码头。

The Conservatory House in Varna

The Conservatory House (a.k.a. Home Tree House) is designed for enjoyment of nature and music. Client asked for a residence with a space for hosting small music events and a large conservatory for growing flowers. These requirements were in line with local building code for rural areas which required a collateral agricultural use for granting a building permit.

Steep and picturesque but compromised site used to be a local sand quarry for the neighboring village and turned into an eroded waste dump later. Serenity of surrounding nature inspired architects to work for restoring the initial equilibrium on site. Resulting design solution fit the house into the existing quarry pit and made the new structure as compact as possible.

Reinforced concrete structure is chosen because of its affordability and local popularity. It is designed with a central core and load-bearing facade frames without internal columns and shear walls. Diagonal, vertical and horizontal structural elements on the southern and eastern facades follow the structural stress lines and reveal the building's tectonics. Lack of cultivated landscaping promotes re-growth of local plant species and preserves local microclimate. Bio-active wastewater treatment unit turns waste into bio-compost and irrigation water. Clean agricultural produce grown on-site adds to the green experience of the Conservatory House.

温室住宅（又名家庭树房）由Ignatov建筑事务所为热爱自然和音乐的房主量身打造。主人要求住宅必须可以承接小型音乐集会，并可容纳一个大型花草温室。这些要求恰巧符合当地的建筑规定，该规定提到：在乡村地区授予任何建筑许可时，必须以某种农业用途作为附属条件。

住宅所在地地势陡峭、风景如画同时不乏舒适，这里曾是邻村的采砂场，后遭废弃，成为垃圾场。住宅周围静谧的环境激发建筑师的灵感，试图通过设计重拾这里最初的那份平衡。本案最终的设计方案是：将住宅自然和谐地融入到原有的采砂场中，其他新增结构则要尽可能做到紧凑。

因为价格合理，又在当地十分普及，建筑师选择钢筋混凝土作为主要结构材料。整个建筑结构只包含一个中立柱和承重的立面框架，没有其他任何内柱或者抗震墙。东南两个立面的所有（包括倾斜、垂直和水平方向上的）结构要素均遵循建筑的结构应力线，展现出建筑的总体构造。由于住宅周围鲜有种植物，因此，必须采取有效措施促进当地植物物种的重新生长，保护当地小气候不受影响。生物活性废水处理装置将废物转变成生物堆肥和灌溉用水。周围农作物更为温室住宅平添绿色体验。

Name of Project / 项目名称:
The Conservatory House
Location / 地点:
Varna, Bulgaria
Area / 占地面积:
780 m²
Completion Date / 竣工时间:
2010
Architecture / 建筑设计:
Ignatov Architects
Landscape / 景观设计:
Ignatov Architects
Interior Design / 室内设计:
Ignatov Architects
Photography / 摄影:
Ignatov Architects

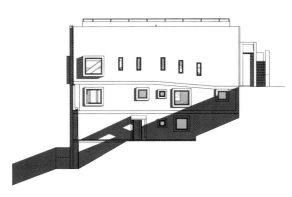
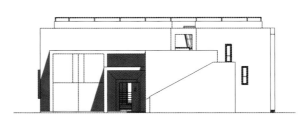

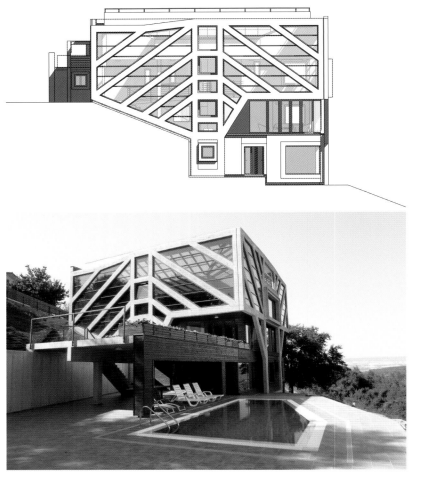

The house filled the void and became a retaining wall itself which allowed restoration of the terrain around it. Conservatory and music room naturally merged together and were placed on top of the residence for catching sunlight and views and minimizing building footprint. The carbon footprint in turn is reduced by the insulating effect of the conservatory over the residence and by utilizing a complex geothermal system that covers all heating and cooling needs. It cleanly and quietly exchanges thermal energy with earth via six closed-loop probes requiring minimal electricity for circulation only. All domestic hot water is supplied by solar vacuum tubes integrated into the glazed roof. The performance bottom line for the past seasons shows that the Conservatory House provides great comfort and uses very little external power without being an encapsulated passive house. In contrary, its concept actively promotes human interaction with seasons and elements.

房子填补了空地，并形成一面挡土墙，保护周围地形不被破坏。温室和音乐室两个空间融合自然，被设在房屋顶层以获取充足的光照和开阔的视野，同时使建筑碳足迹最小化。反之，温室绝缘效应也有助于减少碳足迹。住宅装有一个综合地热系统，可满足房内所有供暖和制冷需求。地热系统污染少、噪声低，六个闭环探针仅需少量用于循环的电量就可以实现地源热能的利用。本案所有家用热水均由集成在琉璃屋顶上的太阳能真空管提供。过去几个季节的居住体验证明，住宅所需的外部电源很少，并且极为舒适。同时，与常见的密封被动式住宅不同，温室住宅积极鼓励人与自然的互动。

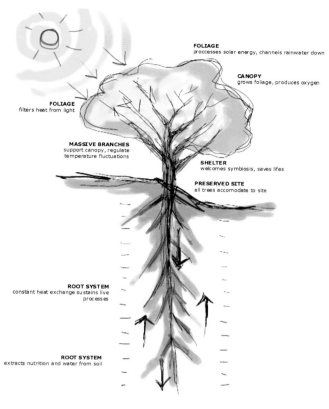

LOCAL WALLNUT TREE

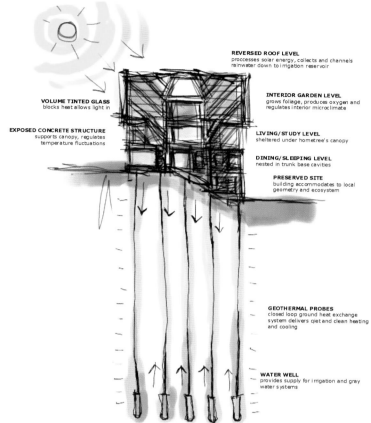

HOME TREE CONCEPT

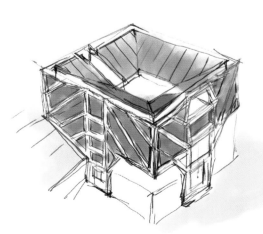

IN GREEN ! RESIDENT ARCHITECTURE • HOUSE

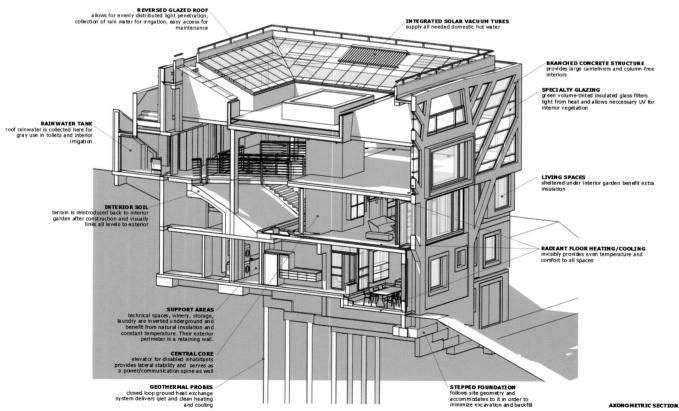

REVERSED GLAZED ROOF
allows for evenly distributed light penetration, collection of rain water for irrigation, easy access for maintenance

INTEGRATED SOLAR VACUUM TUBES
supply all needed domestic hot water

BRANCHED CONCRETE STRUCTURE
provides large cantilevers and column-free interiors

SPECIALTY GLAZING
green volume-tinted insulated glass filters light from heat and allows neccessary UV for interior vegetation

RAINWATER TANK
roof rainwater is collected here for gray use in toilets and interior irrigation

LIVING SPACES
sheltered under interior garden benefit extra insulation

INTERIOR SOIL
terrain is reintroduced back to interior garden after construction and visually links all levels to exterior

RADIANT FLOOR HEATING/COOLING
invisibly provides even temperature and comfort to all spaces

SUPPORT AREAS
technical spaces, winery, storage, laundry are inserted underground and benefit from natural insulation and constant temperature. Their exterior perimeter is a retaining wall.

CENTRAL CORE
elevator for disabled inhabitants provides lateral stability and serves as a power/communication spine as well

GEOTHERMAL PROBES
closed loop ground heat exchange system delivers qiet and clean heating and cooling

STEPPED FOUNDATION
follows site geometry and accommodates to it in order to minimize excavation and backfill

AXONOMETRIC SECTION

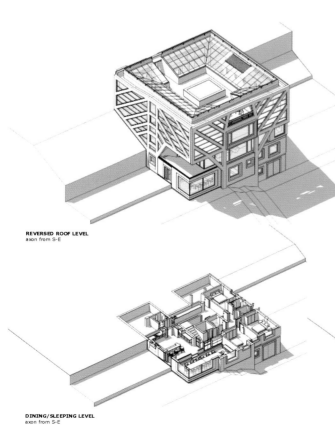

REVERSED ROOF LEVEL
axon from S-E

DINING/SLEEPING LEVEL
axon from S-E

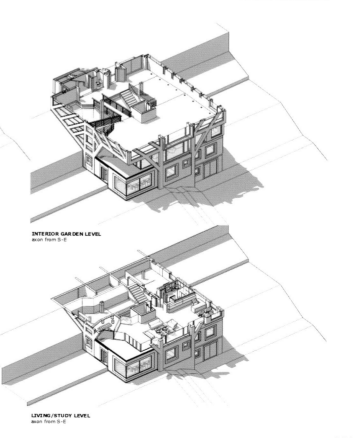

INTERIOR GARDEN LEVEL
axon from S-E

LIVING/STUDY LEVEL
axon from S-E

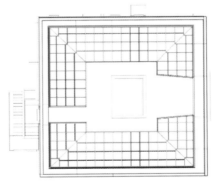

REVERSED ROOF LEVEL
plan

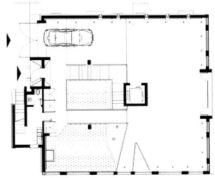

INTERIOR GARDEN LEVEL
plan

DINING/SLEEPING LEVEL
plan

LIVING/STUDY LEVEL
plan

The Houl in UK

The house is sited in the natural concave area of a hillside which faces principally west along the contours to enjoy the spectacular landscape setting of the river Ken valley and the ridges of the Rhinns of Kells hills opposite.

The intention was to create a contemporary single-story "long house" which is recessive in the landscape, sustainable in its construction, very low in energy consumption, and aiming for zero net emissions of carbon dioxide for all energy use in the house.

The entrance to the house is sited on the north east side of the house under the cover of the roof to provide shelter from the prevailing wind. The principal rooms are situated along the contour of the site to enjoy the views across the valley to the west. The ancillary spaces are generally to the rear.

The house is net "zero carbon" by using very high levels of insulation, minimising air infiltration heating, using an air source heat pump with a whole house heat recovery ventilation system, and generating electricity using a wind turbine.

The Houl住宅位于山坡上一处天然凹陷区域，方向大体朝西，从这里可以尽情欣赏Ken河谷和其对面Rhinns of Kells山脉壮丽的自然景观。

建筑师意图打造一个现代的单层"长屋"，让其隐蔽在所处的迷人景观中。建筑是可持续性的，能源消耗很低，更以房内能源消耗实现零碳排放量为努力目标。

出口位于建筑的东北侧，其上方的屋顶有效地阻挡了盛行风的侵袭。房内主要功能区方位基本同其所处地形的等高线持衡，方便房主欣赏河谷美景。其他空间则大多被设在住宅的后方。

多种可持续性策略的使用令The Houl实现了零碳排量的目标，这些策略包括：采用高绝缘度的材料、减少空气渗透、使用空气源热泵并安装整体热回收通风系统，以及利用风力涡轮机产生电能等。

Name of Project / 项目名称:
The Houl
Location / 地点:
Scotland, UK
Area / 占地面积:
185 m²
Completion Date / 竣工时间:
2010
Architecture / 建筑设计:
Simon Winstanley Architects
Landscape / 景观设计:
Paterson Landscape
Interior Design / 室内设计:
Simon Winstanley Architects
Photography / 摄影:
Andrew Lee
Client / 客户:
Simon Winstanley

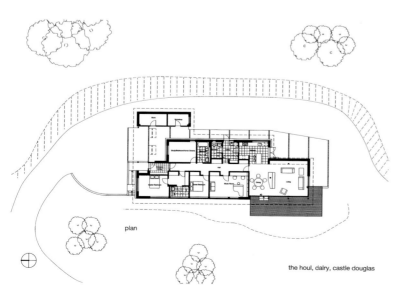

plan

the houl, dalry, castle douglas

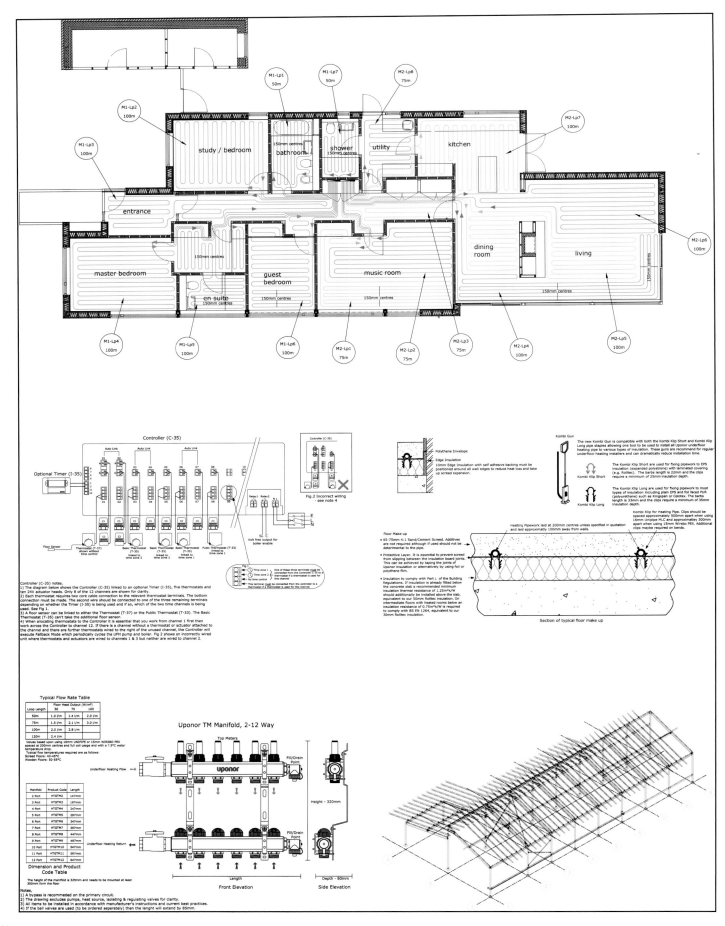

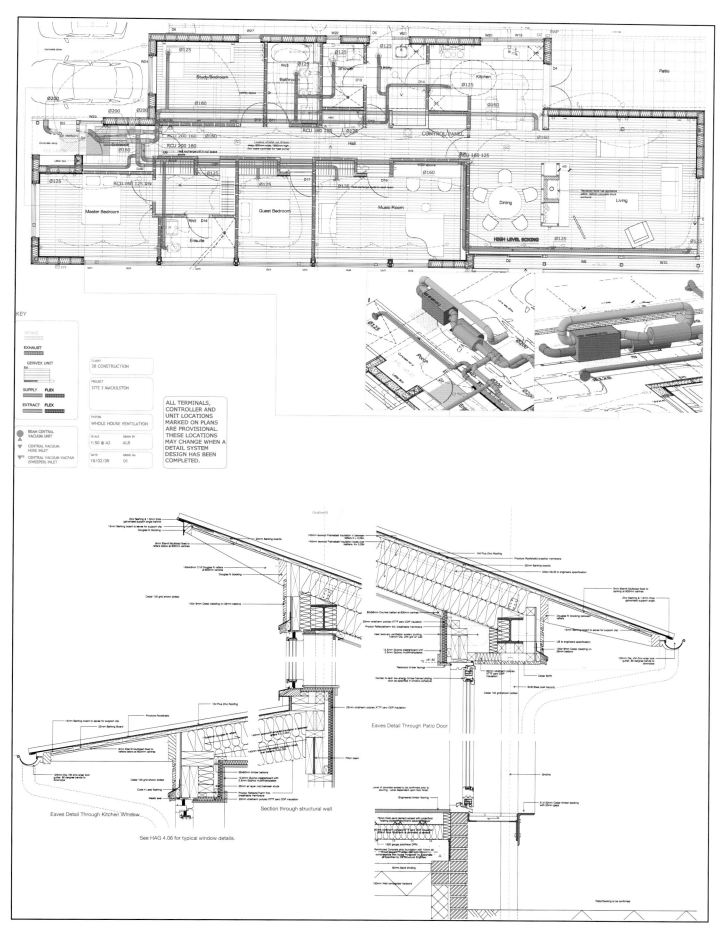

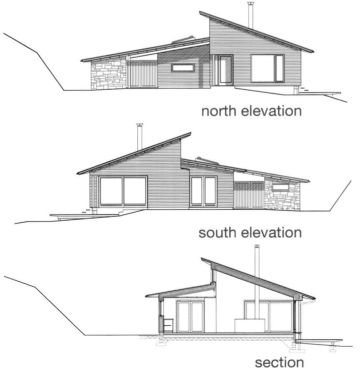

north elevation

south elevation

section

The design uses lightweight but highly insulated steel and timber frame construction, clad in cedar weatherboarding allowed to weather to a natural silver grey colour. The roof finish is pre-weathered grey standing seam zinc. The windows and external doors are triple glazed high performance timber, and were painted grey colour. All insulation levels are to Passiv Haus standards.

The slope of the roof of the main living accommodation follows the slope of the hillside, with the rear roof meeting the main roof at a shallower angle to allow morning sunlight to penetrate the centre of the house.

住宅采用轻型高绝缘度的钢材和木材框架结构，上面覆盖着雪松护墙板，这些护墙板可以被逐渐风化成自然的银灰色。屋顶采用灰色防风化立缝式锌板饰面。住宅窗户和外门材料是装有三层玻璃的高性能木材，被刷成灰色。住宅内所有材料的绝缘水平均达到Passiv Haus标准。

主要居住空间的屋顶坡度同山坡的自然坡度基本吻合，住宅后部的屋顶同总屋顶形成小的角度，利于早晨的阳光照进房子的中心区。

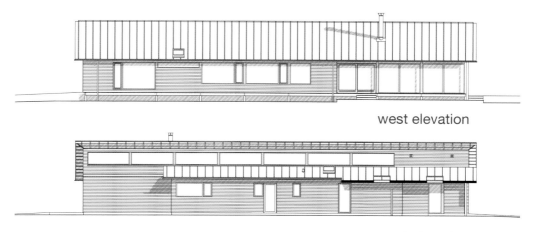

west elevation

east elevation

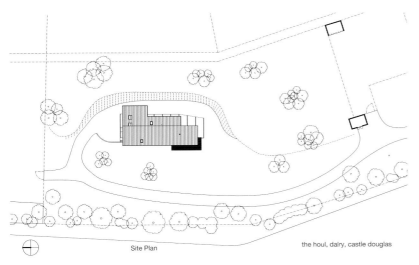

Site Plan

the houl, dairy, castle douglas

The Rent House in Moscow

This house is located forested shores of the Istrinsky reservoir where kept the protogenic beauty of the near Moscow nature and demand solicitous attitude to them.

The house rooms are small and compact, but the lay-outs are very functional. The cottage is not only attractive outwardly, but also warm, reliable, faultless from the ecological point of view and very comfortable for its inhabitants.

The fencing of the gallery and the terrace finishing it is made of wooden laths, with small gaps left between them. As a result while being inside, the architects get the effect of "transparent" wood. Big windows and the complete glass coverage of one of the facades provide the dialogue of internal space with the outside nature. Only one facade is absolutely blind; it is covered with "puzzles" compiled from the rests of boards and laths and toned in different shades of wood.

本案位于Istrinsky水库旁草木丛生的岸边。这里保留了莫斯科附近自然景观的原生之美，也体现了人们崇尚自然的热切态度。

房内房间虽小且布局紧凑，但却具有非常实用的功能性。别墅不仅外观颇具吸引力，整体氛围更是温馨，给人踏实安全的感觉，从居住舒适性和生态角度考虑也都是无懈可击的。

露台及走廊四周由木板条围成，彼此之间留有空隙。这种设计使人们走进去时，便会感受到"透明"木的效果。巨大的玻璃窗和一面完整的玻璃外墙使室内空间与室外自然风景交相辉映。建筑中只有一个外立面是完全封闭的，上面由明暗色调不同的木板条拼接而成，呈现拼图式图案。

Name of Project / 项目名称:
The Rent House
Location / 地点:
Moscow, Russia
Area / 占地面积:
206 m²
Completion Date / 竣工时间:
2010
Architecture / 建筑设计:
Totan Kuzaembaev Architectural Workshop,
Kuzembaev Totan,
Safiullin Danir,
Chertkova Alexandra
Landscape / 景观设计:
Totan Kuzaembaev Architectural Workshop
Interior Design / 室内设计:
Client
Photography / 摄影:
"Individual House" Company

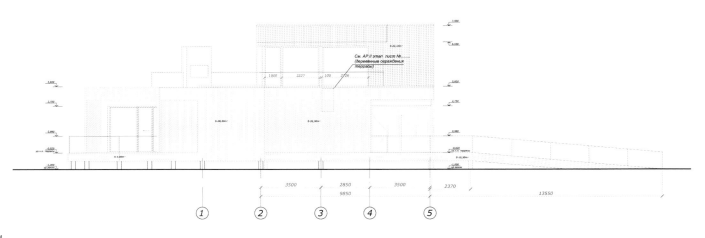

255

The task of the architect was not to harm the surrounding nature and to keep the sense of unity with it. For this reason timber has been chosen as a material for cottages in the territory of the resort "Lazurny Bereg" ("French riviera" or the Cote d'Azur) – one of the warmest, human-friendliest and closest to the nature material.

The cottage is built within the timber framing technique, hence it is rather lightweight. In order to make this construction feature be sensed visually the architect put it on the pier foundation. The house placed above the earth level provides minimum intervention in the existing landscape. On the flat roof of the one-storied space there is a big terrace a part of which is covered with greenery. The terrace railing is made of glass not to hide the nice view which opens from there.

From the outside the house is finished with larch boards. The boarding is made according to the system of ventilated facades. To prevent heat loss the insulators used are 2 times thicker than necessary according to the Russian building standards. The southern facade of the house is as glazed as possible which allows to use the sun warmth in winter. And in summer when the sun is high the canopy provides shade.

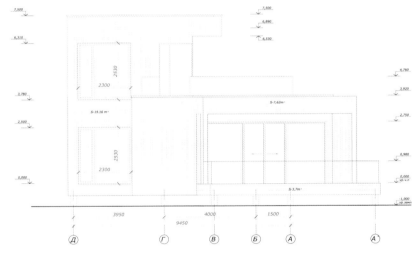

在建筑过程中，建筑师关注周围自然环境，使其不受破坏，并将建筑项目与自然环境完美融合。因此，建筑师选用木材作为度假别墅Lazurny Bereg的建筑材料。木材是最为温和、对人类最有益的天然材料之一。

别墅以木质框架结构为主，因此具有重量较轻的特性。为使建筑在视觉上产生震撼的效果，建筑师将本案设置在墩式地基之上。将别墅设置在地平面以上的方法，将会对自然景观造成最小的影响。在这仅有一层空间的平屋顶上，一个巨大的露台被建于其上。露台的围栏由玻璃制成，可以便于欣赏自然风景。

建筑外观选材落叶松木。留出空隙的木板建筑立面设计利于通风。根据俄罗斯建筑标准，为防止热量损失，本案使用的绝缘体比普通建筑厚两倍。南向立面装有巨大明亮的玻璃窗，便于冬季吸取自然光线。夏日，艳阳高照时，顶棚还可以遮住阳光，保持室内凉爽。

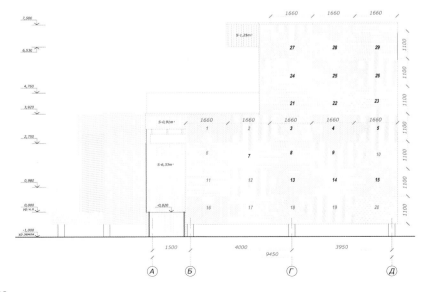

257

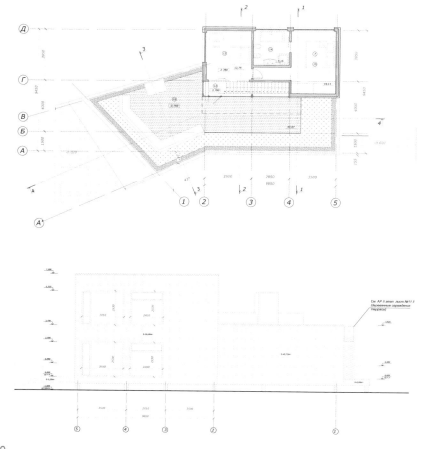

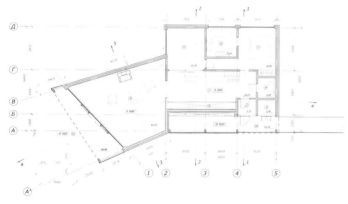

Villa BH in the Netherlands

Villa BH is a modern, environment-friendly house with a remarkable experience of space, light and natural context. The villa is positioned on a rectangular plot of 35 x 50m, that is enclosed at 3 sides with similar plots and freestanding houses. On the back (northeast) of the plot there's an old embankment with several tall trees. From the living program, the kitchen, dining area and living are all orientated on this green scenery. Here the villa has a glass façade over 20 meters long.

Villa BH is inhabited by a couple about 60 years old. To optimize the accessibility of the house, all the program is situated on the ground floor around a patio. This enclosed outdoor space provides the owners the privacy they admire. At the same time, the patio makes the living area an enlightened space and gives it a façade to the south. The floor plan is very open and a concatenation of program. The specific form of the patio widens and narrows the interior space, making it a variety of areas. The façade of the patio is completely from glass panels, giving the villa great perspectives in its interior but also towards the context.

From the main bedroom, which is opposite of the living, there is a layered see through towards the existing embankment with the several tall trees as a central focus point on the plot. The ceiling of the living area has an extra height in the shape of a sloped roof. The physical appearance of this area is very unique and highly qualitative. Lifting the roof in this area enlarges the perspectives, from all the different areas inside villa BH to the existing treetops, which give this plot it's specific character.

Villa BH是一座时尚、环境友好型别墅建筑，有着超乎想象的空间感、自然光线及自然环境。本案位于长50米、宽35米的矩形地块上，其三面被相似结构的独立式房屋所环绕。东北方向的路堤旁高耸着翠绿的树木。在建筑内部，从厨房、餐厅、客厅向外望去，可以欣赏到户外的绿色风景。建筑玻璃幕墙长达20米。

别墅的主人是一对大约60岁的夫妇。为使主人的日常生活更加方便，室内的各项日常物品以露台为中心，安置在一层。户外空间为主人营造出他们所青睐的私密生活氛围。客厅坐北朝南，露台的布局令该空间更显明亮。开放式格局将各个部分连为一体。或宽或窄的露台形式变化令室内空间也随之发生改变，特色鲜明。露台正面完全由玻璃幕墙围合而成，使室内空间与户外风景互为通透，视野十分开阔。

在客厅对面，从主卧房向绿树丛生的路堤方向望去，视觉层次分明，形成一处景观焦点。居住区的天花板采用倾斜式屋顶的构造方式，令室内高度备感增加。该区域的物理外观是非常独特且高度定性的，将屋顶抬高的设计方式扩大了视角。从别墅内各个不同区域到户外的树梢都将成为这里景观的一角，充分展现本案独特的设计风格。

Name of Project / 项目名称:
Villa BH
Location / 地点:
Burgh-Haamstede, the Netherlands
Area / 占地面积:
1.751 m²
Completion Date / 竣工时间:
2010
Architecture / 建筑设计:
WHIM Architecture
Landscape / 景观设计:
WHIM Architecture
Interior Design / 室内设计:
WHIM Architecture
Photography / 摄影:
Sylvia Alonso

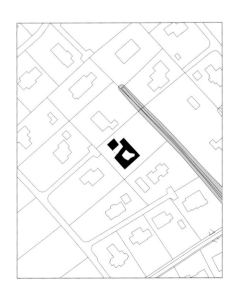

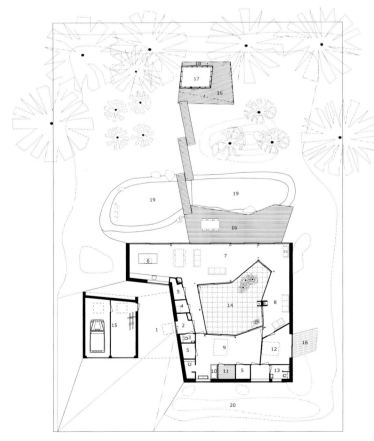

1. Carport
2. Entrance
3. Toilet
4. Installation room heat pump
5. Closet
6. Kitchen
7. Living
8. TV-room
9. Main bedroom
10. Bath
11. Sauna
12. Guestroom
13. Guest bathroom
14. Patio
15. Garage
16. Terrace
17. Garden house
18. Wood storage for fireplaces
19. Pond
20. Wadi

The villa is designed as environmental-friendly with extra insulated façades, roofs and floors. The roof is covered with sedum, which regulates the distribution of the rainwater gently. On the flat roof are 20 solar panels for electricity. A heat pump warms the floors in the winter and cools them in the summer with natural temperature differences retrieved deep in the ground. As an extra heating, there are 2 fireplaces for wood, one in the living and one in the TV-room.

设计师将本案定位为具有高度绝缘性房屋立面、屋顶及地面的环境友好型别墅。屋顶覆盖着的景天属植物，可以温和地调节雨水的分布。屋顶平台上的20个太阳能电池板，为生活提供电力能源供给。热泵通过地下自然温度调节，在冬日里使室内更加温暖，在夏日里则令室内更为凉爽。置于客厅及电视房内的两个壁炉可以通过燃烧木材为室内提供更多的热量。

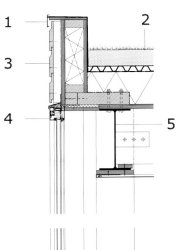

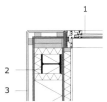
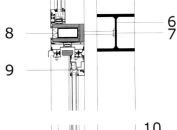
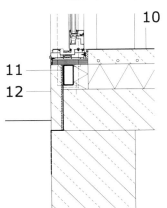

Vertical detail North-East facade

1. Sheet aluminium covering
2. Roof construction:
   .Substrate layer 80mm
   .Filter mat
   .Drainage 25mm
   .protective layer
   .Waterproof foil
   .Insulation 160mm
   .Vapour-retarding layer
   .Plywood 22mm
   .Structural beams 96x271
   .Acoustic ceiling
3. Siberian Larch tongued-and-grooved, Waxed
4. Aluminium window frame with **double** insulated glass
5. Steel beam IPE 300
6. Steel beam HEB 180
7. Steel connection 15x15
8. Steel tube 100x50
9. Aluminium sliding door with double insulated glass
10. Floor:
    .Synthetic seamless floor finishing 3mm
    .70mm screed around underfloor heating
    .Insulation 120mm
    .Reinforced concrete slab 200mm
11. Steel tube 50x100
12. Prefab concrete

Horizontal detail corner North-East facade

1. Aluminium sliding door with double insulated glass
2. Structural steel column, HEB 180
3. Wall:
   .Siberian Larch tongued-and-grooved, Waxed
   .Cavity 38mm
   .Waterproof vapour transm. foil
   .Plywood 18mm
   .Insulation + wood construction 246mm
   .Plywood 18mm
   .Insulation 59mm
   .Vapour-retarding layer
   .Plasterboard 15mm

IN GREEN ! RESIDENT ARCHITECTURE • HOUSE

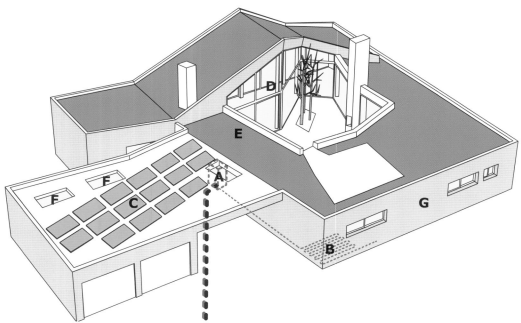

3D VOLUME Sustainability

A. Heat pump
B. Floor heating/cooling
C. Solar panels
D. Glass facade to the South
E. Sedum roof
F. Skylight
G. Timber facade

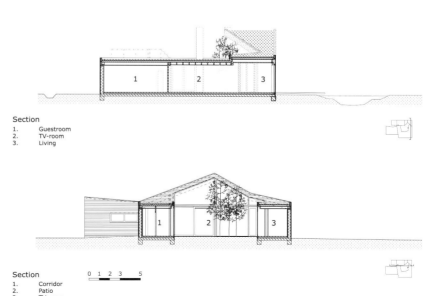

Section
1. Guestroom
2. TV-room
3. Living

Section
1. Corridor
2. Patio
3. TV-room

0 1 2 3 5

Villa Nyberg in Central Sweden

Göteborg-based Kjellgren Kaminsky has produced Swedens first series of passive houses sold as type houses in collaboration with Emrahus. Their goal is to make this environmentally friendly building technique available for all. Villa Nyberg is the first one to get built. The villa has been customized for the Nyberg family and is situated in Borlänge, central Sweden. Designed as a round structure in order to minimize materials and increase efficiency, the home is so efficient that it even blows away some of the passivhaus standards that it was designed for!

The prefab home is 156 square metres with a complete circular lower floor and an almost half circle space on the second floor. A central atrium contains the staircase to the second floor and draws in light to the inner core of the house. The villa is situated by a lake in a forest in central Sweden. The living room and kitchen open up to views of the lake, while the more private bedroom and bathroom with smaller windows overlooking the forest are located in the back. The round shape of the villa reduces the enclosing wallarea of the house. It also effects the way one lives in the house. During the day, one will move from room to room around the building, experiencing different views and daylight conditions.

位于哥德堡的Kjellgren Kaminsky Architecture与客户Emrahus合作，将在瑞典设计的首批被动节能房屋以标准房屋形式出售。他们的目标是令环境友好型建筑技术可以得到广泛应用。Villa Nyberg成为第一所被建房屋。该别墅位于瑞典中部城市博朗厄，是专为Nyberg一家精心设计的。建筑师将本案设计为圆形结构，以期减少材料的使用，同时增加能效。最终该设计远远超越了原本制定的被动节能标准。

这个预制建筑占地156平方米，一层是圆形空间，二层则是半圆形空间。中央客厅内的楼梯通向楼上二层，室内中心区域充满自然光线。本案位于瑞典中部森林湖边，客厅、厨房面对小湖而建，更加私密的卧室及设有较小窗户的浴室则被设在房屋后面，可以俯瞰森林。别墅的圆形布局设计减小了房子的封闭墙体区域，同时也对日常生活产生影响。白天时，人们可以到环绕别墅周围的不同房间中，从不同角度去欣赏风景及享受光照。

Name of Project / 项目名称:
Villa Nyberg
Location / 地点:
Borlänge, Sweden
Area / 占地面积:
156 m²
Completion Date / 竣工时间:
2009
Architecture / 建筑设计:
Kjellgren Kaminsky Architecture
Landscape / 景观设计:
Kjellgren Kaminsky Architecture
Interior Design / 室内设计:
Kjellgren Kaminsky Architecture
Photography / 摄影:
Kalle Sanner
Client / 客户:
Emrahus & the Nyberg Family

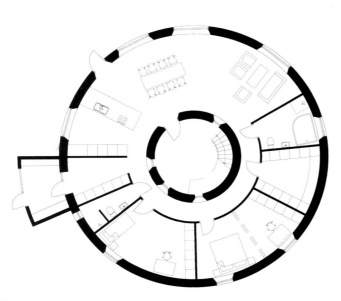

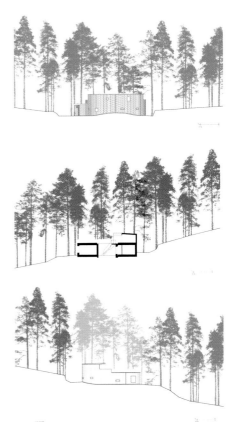

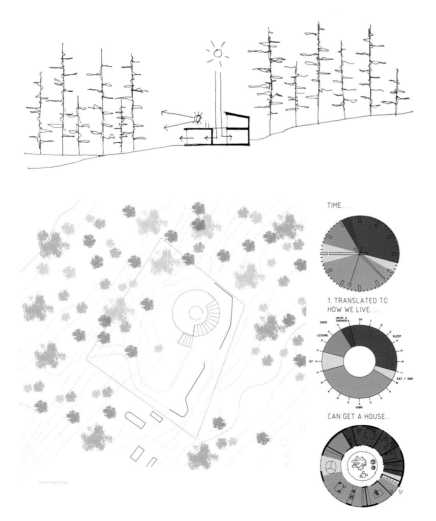

Built to passivhaus standards, Villa Nyberg is incredibly energy efficient and includes super-tight insulation. Passive houses are extremely well-insulated buildings that are largely heated by the energy already present in the building – people and the household equipment generate a lot of energy. Heating for the home is largely provided by excess heat from inside the home, which is trapped inside by the tight envelope. In fact, the home is so insulated that it actually broke a Swedish record, and energy simulations show that this villa will have an energy consumption for heating of only 25 kWh/m² per year. On top of all this, a solar hot water heater on the roof provides hot water for the home.

To reduce heatloss airtightness is an important feature of a passive house. When they tested the airtightness of Villa Nyberg, they got a pleasant surprise, a new Swedish record! The result was 0.038 L/m² at 50 Pa. The Swedish passive house standard is 0.3 and the previous record was 0.07.

以被动式节能标准为目标，Villa Nyberg在密封绝缘乃至整体节能方面达到了惊人的效果。被动式节能房要有良好的隔热性，其热量主要源于建筑中的已存能源——人们及家庭设备产生的诸多能量。该房供暖主要源自建筑内部产生的剩余热量，这些热量由于房屋密封良好而得以保存。实际上，房屋极佳的绝缘性能已经打破瑞典记录，能量模拟结果表明该别墅在热能方面的消耗仅仅是25千瓦时/年·平方米。此外，屋顶上为人们提供日用热水的太阳能热水器起到至关重要的节能作用。

为减少热量损失，密封性能成为被动式节能建筑的一项重要指标。该别墅的密封性能经测试达到令人叹为的效果，在50帕气压下以0.038升/平方米的指标开创瑞典记录，瑞典被动节能建筑标准指标为0.3升/平方米，之前记录为0.07升/平方米。

White Shade in Australia

With an area of 203 square metres, White Shade was designed by Calder Flower Architects. Located in Sydney, Australia, the project was completed in 2010.

White Shade is a simple new house on a flat site on Sydney's Lower North Shore. As lovers of modernism, the clients wanted an elegant, energy-efficient home robust enough to withstand the rigours of day-to-day family life.

The house has been conceived and designed to be an elegant and practical energy efficient home. The long axis of the house faces due north with overhangs calculated to admit winter sun and shade in summer. Windows and doors are double-glazed. The house combines concrete frame structure with reverse masonry veneer infill panels to enhance thermal and acoustic performance. The concrete is left exposed. The cabinetry, fenestration and stair are expressed as insertions in the concrete frame. Rain water storage is located under the floor slab with in-slab hydronic heating.

White Shade由CFA建筑事物所担纲设计，占地203平方米。本案位于澳大利亚悉尼市，于2010年竣工。

简约崭新的家居建筑White Shade位于悉尼北海岸的平坦区域。作为现代主义风格的衷爱者，业主意在打造一座高雅、节能的家居建筑，使其节能强度足以承受日常家庭生活所需。

最终，该房被成功打造成高雅、实用的高效节能建筑。长长的房屋轴线面向正北，屋檐的设计使冬季采光和夏季遮阳的功效兼备。门窗都装有双层玻璃。混凝土框架和反向砌石贴面构成建筑主体，以期增强保暖效果及隔音效果。混凝土材料被暴露在外。细工家具、门窗、楼梯被设计成嵌入混凝土框架中的形式。雨水储存于楼板以下，楼板内设有液体循环加热系统。

Name of Project / 项目名称:
White Shade
Location / 地点:
Sydney, Australia
Area / 占地面积:
203 m²
Completion Date / 竣工时间:
2010
Architecture / 建筑设计:
Calder Flower Architects
Landscape / 景观设计:
Anthony Wyer landscape Design
Interior Design / 室内设计:
Calder Flower Architects
Photography / 摄影:
Hungyu Ko

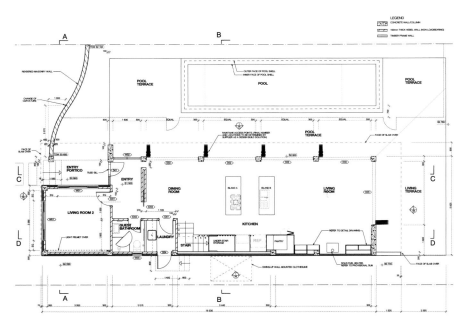

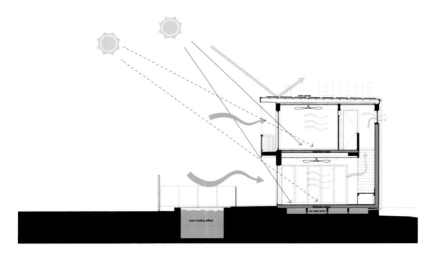

The house has a long narrow footprint along the east-west axis and every habitable room is oriented towards the north and the optimal solar aspect. It has deep overhangs on the north and deeper overhangs to the east and west to shade the windows in the summer months but is designed to allow the low angled winter sun to reach the full interior of the house.

The house is built primarily of concrete. The floors (ground and first) are concrete and majority of the ground floor structure is exposed off shutter concrete. This gives the house significant thermal mass and the ability to hold the warmth from the winter sun. The ground floor concrete slab is polished and exposed as a natural concrete finish. The windows and glazed exterior doors are double-glazed for optimal thermal insulation. The ground floor concrete slab has hydronic heating cast into the slab. There is no air-conditioning in the house. Ceiling fans in every room and windows on the north and south encourage effective cross ventilation which keeps the interior cool in summer. Hot water radiators are installed in the bedrooms. The water for the hydronic heating system is heated by a low energy use gas powered furnace.

There is a 10,000 litre rainwater tank system under the ground floor slab which harvests and stores all rainwater captured from the roof for re-use. Rainwater is used for flushing the toilets, irrigating the garden and maintaining the outdoor lap pool. The garden is comprised of low water use planting. Pebbles and rocks combine in the landscape to protect the soil from exposure and evaporation.

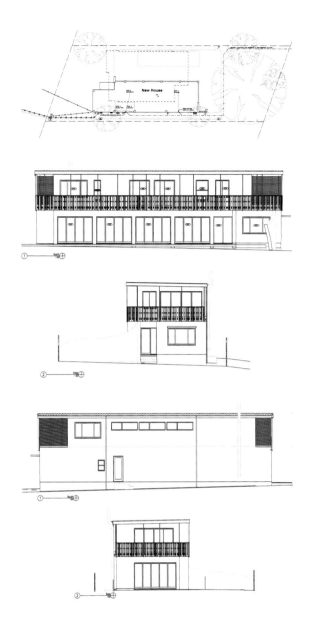

本案为狭长式东西方向建筑，所有可居住的房间都面向北部（本案位于南半球），以实现最佳采光效果。房屋北部设有较深的屋檐，东、西部屋檐深度有所增加，以保证在夏日里遮挡阳光，在冬日里令整个室内空间充满自然光线。

混凝土是本案设计中的主要选材。一层和二层采用混凝土建筑材料，尤其一层框架大部分都是暴露在外的混凝土。该材料的运用使房屋隔温效果良好，保存冬日里阳光的温度。一层的混凝土板经过打磨，形成自然混凝土贴面。窗户和玻璃外门均被装以双层玻璃，更利于隔温。一层混凝土板内设有液体循环加热系统。室内没有空调系统设备。每个房间的天花吊扇及南北窗户都保证了室内通风良好，保证夏日室内凉爽。卧室内装有热水散热器。热水供暖系统的水由低能耗的燃气壁炉加热。

容量为1万升的地下雨水收集系统存储屋顶雨水以便再次利用。收集而来的雨水可用来冲洗卫生间、浇灌花园、供给户外小型水池。花园内种植着用水量较少的植物。户外景观中使用的鹅卵石和岩石防止土壤暴露，利于维持土壤湿度。

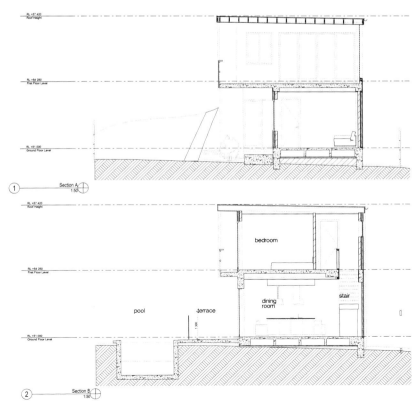

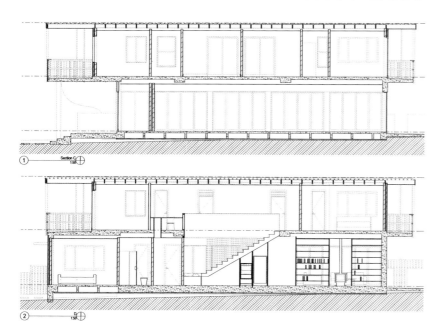

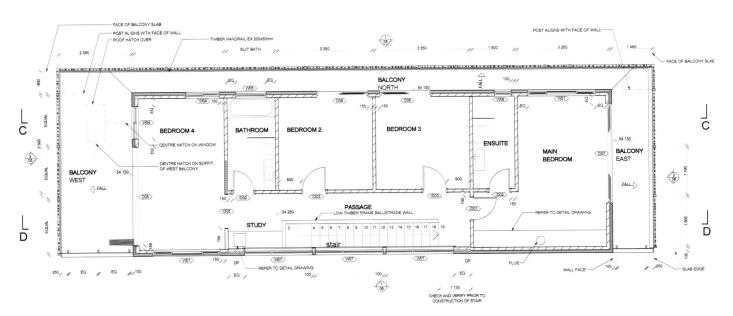

282-311
APARTMENT
RESIDENT VOLUME
PART 3

Da Vinci Residential Tower

The project, which is planned as a condominium apartment tower vertical, consists of 38 apartments spread over 19 floors, stairwell and three elevators (one of the service) that connect the units with parking. The apartments are located from floor 1 to 19 and are willing only two per floor. They consist of a service area with vertical circulation, with stairs and elevator service, in addition to the main elevator lobbies downloaded distribution to the apartments.

On the ground floor is located the main entrance to the tower, as well as vehicular access to private parking and parking for visitors and people with physical limitations. In total there are 118 parking spaces in basements and 12 more on access level. Since access to both main and service, are located the main lobby and service, elevators and common areas. In general, these consist of a meeting room for condominiums, with kitchenettes and bathrooms, a lounge, and a spa with swimming lanes, jacuzzi, locker rooms and bathrooms. In addition, the tower has paddle tennis court and pool on the roof exposed to sun decks, bar and recreational areas. Beyond the Da Vinci Tower with gardens, children's areas and other areas.

建筑师意在将本案打造为19层的公寓大楼，内部设有38户公寓房、楼梯间及将各单元连接起来的三部电梯（公寓内服务项目之一）。从底层到顶层，每个楼层只容纳两户。楼梯间和电梯间构成了公寓大楼内垂直分布的循环服务区。除了主电梯大堂外，公寓各楼层均有分布。

公寓一楼的主入口可以通向公寓内部，同时设有私人停车场和为来访者及残疾人准备的停车专用区。停车场共有118个地下停车位及12个备用车位。建筑师为公寓大楼设计了一间设有小厨房及卫生间的会议室，一间休息室，带有泳道、按摩浴缸、更衣室和浴室的SPA温泉浴场。除此之外，屋顶设有网球场、露天泳池、阳光甲板、酒吧及休闲区。儿童区和其他区域以及花园也都应有尽有。

Name of Project / 项目名称:
Da Vinci Residential Tower
Location / 地点:
Edo de Mexico, Mexico
Area / 占地面积:
16,650 m²
Completion Date / 竣工时间:
2007
Architecture / 建筑设计:
Pascal Arquitectos
Photography / 摄影:
Sofocles Hernandez
Client / 客户:
Grupo Inmobiliario Rofessi S.A. de C.V.

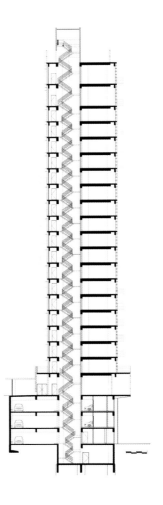

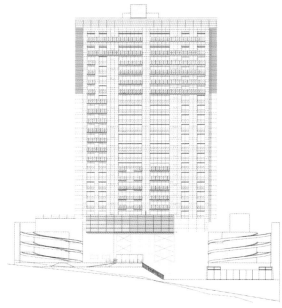

The whole facade is ventilated and cladding elements are hung, screwed or stapled, making it completely detachable. A thermo-meteorological acoustic barrier was created and in consequence, this is very energy efficient building. At the same time, this system facilitates façade maintenance and replacement, and a convenient access to hydraulic and sanitary installations goes through along the perimeter of the building.

All of the residual water is treated and re-used in garden watering, car wash, and toilets. This building also comprises intelligent systems for the building access, CCTV, energy saving lighting controls and lamps.

Regarding the project's construction system, the building consists of a reinforced concrete structure based on beams, columns and slabs, with two side castings, and a foundation based on concrete piles up to 2 meters in diameter and 22 metres in depth.

整个建筑外观的设计保证室内通风良好，外观饰面元素通过螺丝及钉子加固，悬于外部并与建筑成完全分离的状态。由此形成的热气象隔音屏障，使本案成为高效节能的公寓建筑。除此之外，这一建筑结构还利于公寓外观的保养及更换，同时也便于环绕建筑外周的水力和卫生设备的装置。

所有废水经过处理后可以用来浇灌花园、洗车或者冲洗卫生间。公寓楼内还装有安全智能门禁系统、闭路电视以及节能型的灯光控制设施和节能灯具。

本案的建筑系统包括以横梁、立柱、楼板为主体的钢筋混凝土结构，两侧铸件，以及直径2米、高约22米的混凝土桩地基。

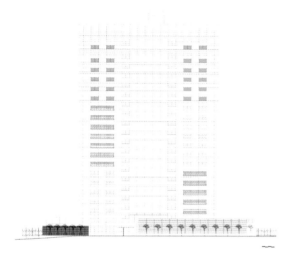
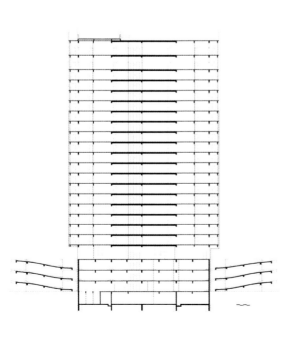

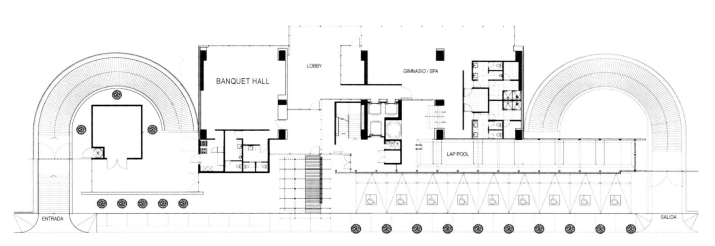

Sierra Bonita Affordable Housing

The Sierra Bonita Mixed-Use Affordable Housing was constructed by the city of West Hollywood to address an affordable housing shortage for tenants living with disabilities. The mixed-use program brings higher density into the urban core of the city. Locating the project within the central urban fabric of the community ensures that residents have direct access to local businesses and services. Multiple public transportation options are directly accessible on the busy transit corridor of Santa Monica Boulevard minimizing the need for private transportation.

Sierra Bonita serves as the pilot project for the city of West Hollywood's green ordinance, which was passed in 2007. The arrangement of the proposed architectural scheme facilitates an environmentally conscious approach to the building services design. Passive solar design strategies are used and include: a north south orientation for the living units, locating and orienting the building to control solar cooling loads, shaping and orienting the building for exposure to prevailing winds, designing windows to maximize daylight, minimizing west-facing glazing and designing units to maximize natural ventilation. A photovoltaic panel system is integrated into the façade and roof of the building that will supply most of the peak load electricity demand. The panels are integral to the building envelope and the unused solar electricity will be delivered to the grid.

Sierra Bonita多用途经济适用住宅由西好莱坞市政府建造，用以缓解该区残障人员用房紧张的问题。该项目被设在该城的核心区，不仅增加了该区的人口密集度，同时保证其住户更方便快捷地享受当地商业和其他配套服务。圣莫尼卡大街是繁华的交通要道，可提供多种便捷的公共交通工具，最大限度地减少了当地居民对于私人交通工具的需求。

Sierra Bonita是西好莱坞市于2007年推行的绿色条例的试点工程。建筑团队所提出的建筑方案中，环境保护策略被有效地融入到建筑设计中去。多种被动式太阳能设计得以采用，其中包括：居住单元呈南北走向，建筑的方位便于控制太阳能系统冷负荷，建筑形态和方向上的设计扩大了房屋对盛行风的接触面积，窗户设计令日光的利用率达到最大化，建筑西侧玻璃使用最少化，使房内自然通风效果最佳。建筑外壁和屋顶安装了光伏板系统，用电高峰期所耗费的大部分电力均由其供给。太阳能光伏板同建筑外观完美结合，余下的太阳电能将被传送到电网。

Name of Project / 项目名称:
Sierra Bonita Affordable Housing
Location / 地点:
West Hollywood, USA
Area / 占地面积:
4645 m²
Completion Date / 竣工时间:
2010
Architecture / 建筑设计:
Patrick Tighe Architecture
Landscape / 景观设计:
AHBE Landscape Design
Photography / 摄影:
Art Grey Photography
Client / 客户:
West Hollywood Community Housing Corporation

ROOF PLAN

FIFTH FLOOR PLAN

FOURTH FLOOR PLAN

THIRD FLOOR PLAN

GROUND FLOOR PLAN

SECOND FLOOR PLAN

SUBTERRANEAN PARKING LEVEL

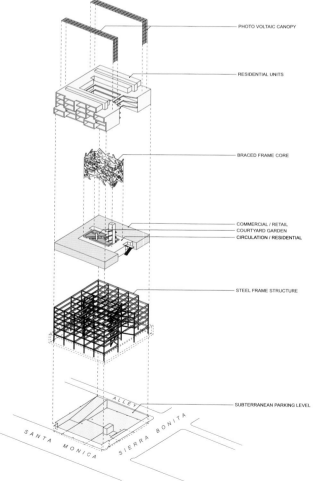

- PHOTO VOLTAIC CANOPY
- RESIDENTIAL UNITS
- BRACED FRAME CORE
- COMMERCIAL / RETAIL
- COURTYARD GARDEN
- CIRCULATION / RESIDENTIAL
- STEEL FRAME STRUCTURE
- SUBTERRANEAN PARKING LEVEL

The green building meets or exceeds all of the green ordinance requirements. This includes multiple facets of green building, such as construction and waste management, storm water diversion, construction air quality, sustainable materials and finishes, water conservation, and energy efficiency.

Specific examples of some of these requirements are the use of energy star appliances, low-VOC interior paint and wood finishes, low-flow showerheads, faucets and water closets, and energy efficient outdoor lighting.

Integral colour stucco and metal siding provide durable exterior finishes minimizing waste associated with higher maintenance exterior envelope materials. Seventy-five percent of the construction and demolition waste will be recycled during construction.

该建筑的各项指标均符合或高于绿色条例的要求，涵盖绿色建筑应当具备的要素，例如，建造及拆卸废物的有效管理、建设空气质量的保护、可持续性材料和处理工艺的利用、水资源以及其他能源保护措施的应用。

绿色条例的一些具体要求包含：节能之星家用电器、低VOC（挥发性有机化合物）室内涂料和木材饰面、小流量淋浴头、水龙头和抽水马桶，以及室外节能照明设备的应用。

彩色灰泥和金属护墙板协调统一，令建筑外观更加持久耐用，同时极大程度地减小了用于高处外墙材料维护所造成的浪费。该建筑在建造和拆卸过程中所产生的百分之七十五的"废物"，都被重新利用在新的建造工程中。

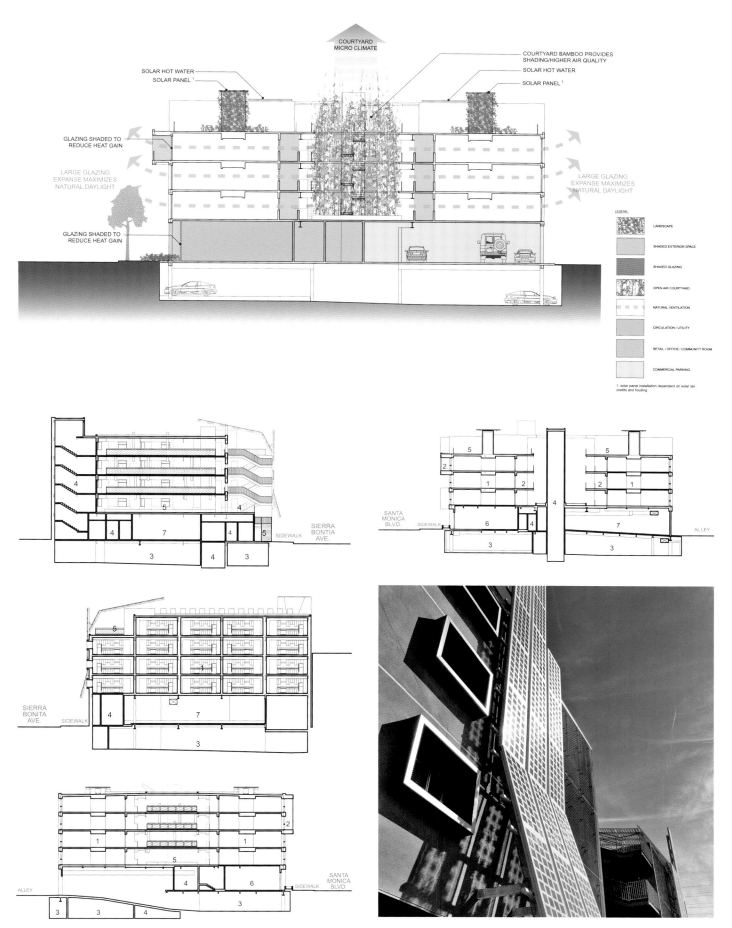

Soe Ker Tie House

In the fall of 2008, TYIN travelled to Noh Bo, a small village on the Thai-Burmese border to design and build houses for Karen refugee children. The 60 year long conflict in Burma has forced several hundreds of thousands of people to flee from their homes. The conflict leaves many children orphaned, with little hope for the future.

A few months earlier, the architects got in touch with Ole Jørgen Edna from Levanger, Norway. Edna started an orphanage in Noh Bo in 2006, and was now in need of more dormitories. From sheltering 24 children, the orphanage would grow to house almost 50. The Soe Ker Tie project was finished in February 2009.

The main driving force behind the project was to somehow recreate what these children would have experienced in a more normal situation. Architects wanted every child to have their own private space, a home to live in and a neighbourhood where they could interact and play. These six sleeping units are the answer to this. Because of their appearance, the buildings were named Soe Ker Tie Hias by the Karen-workers – the Butterfly Houses.

2008年的秋天，TYIN团队前往泰缅边界的一个名为Noh Bo的小村庄，为这些克伦邦难民孩子设计和建造避难住所。缅甸的这场持续了六十年多的冲突已经迫使数十万百姓逃离他们的家园，许多孩子因此成为孤儿，未来渺茫。

几个月之前，TYIN同来自挪威莱旺厄尔的Ole Jorgen Edna女士取得联系。她于2006年在Nohbe创办了一所孤儿院，此时正急需更多的学生宿舍。孤儿院原本只能容纳24个孩子，现已扩大到50人。Soe Ker Tie项目于2009年2月完工。

不论如何要让这些孩子体会相对正常的生活，是隐藏在本案背后的主要驱动力。设计团队希望每个孩子都能拥有属于自己的私人空间，一个属于他们自己的家，一个可以任意玩耍和交流的地方。这六个提供睡觉和休息空间的建筑物，就是建筑团队对于这些孩子也是对于他们自己的答复。因其特殊的外观，这些建筑被当地克伦邦工人称做Soe Ker Tie——像蝴蝶一样的房子。

Name of Project / 项目名称:
Soe Ker Tie House
Location / 地点:
Noh Bo, Tak, Thailand
Completion Date / 竣工时间:
2009
Architecture / 建筑设计:
TYIN tegnestue Architects
Pasi Aalto,
Andreas Grøntvedt Gjertsen,
Yashar Hanstad,
Magnus Henriksen,
Line Ramstad,
Erlend Bauck Sole
Photography / 摄影:
Pasi Aalto
Client / 客户:
Ole Jørgen Edna

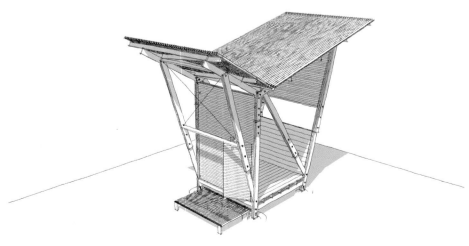

297

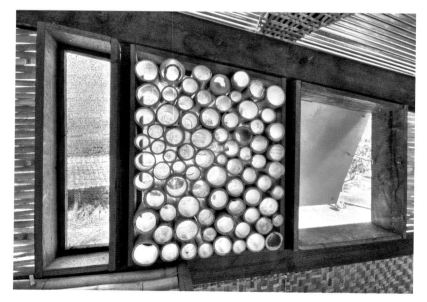
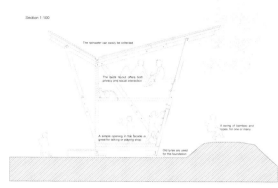
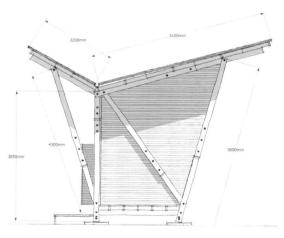

The bamboo weaving technique used on the side and back facades is the same used in local houses and crafts. Most of the bamboo is harvested within a few kilometers of the site. The special roof shape of the Soe Ker Tie Houses enables an effective, natural ventilation, at the same time as it collects the rain water. This renders the areas around the buildings more useful during the rainy season giving the kids better areas for play and social life. The iron wood construction is prefabricated and assembled on-site, using bolts to ensure reasonable precision and strength. By raising the buildings from the ground, on four foundations cast in old tires, problems with moisture and rot in the construction are prevented.

当地建筑和手工艺品中常见的竹编技术被应用在本案建筑两侧和背部的外壁中。屋顶形状特殊，不仅可以促进室内自然通风，同时便于收集雨水。当雨季来临时，建筑物的周围还会形成极为有趣的区域，供孩子们玩耍和交流。硬木构造的建筑物是预制的，在现场组装而成，螺钉的使用将整个建筑物精准牢固地组装在一起。建筑的房基固定在若干旧轮胎上，使房子脱离地面，以避免建筑物遇到潮湿、被腐蚀等问题。

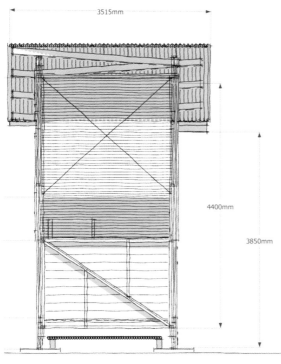

299

301

Urban Lake Housing in Italy

The urban refurbishment of Largo San Giorgio site in Pordenone, Italy follows the main topic of returning the citizens of Pondenone again a part of the town which was inaccessible due to the different levels of the streets and to the fact that it had been private for many years. A path sews together the lake, the park and the square of San Giorgio church designing a public open space where the new housing and the church servces face. As many of C+S works, the architectural design focuses on the strong relationship between architecture and landscape. The skirting board of the buildings is made of Prun stone, the same material the square is made of.

The buildings grow with the levels of the paths surrounding the lake designing a new urban connection which specifically changes material: Prun stone on the square, larch wood on the bridges crossing the lake and ochre concrete on the paths of the park. Detailed architectural design defines also the other public parts of the buildings: entrance halls and public stairs, which are made of red-coloured wood, the auditorium (partially dag in the ground), with grey and red acoustic panels, the church classrooms which are graphically designed, All these spaces are naturally enlighted and look the marvellous landscape of the lake outside. All the houses are opened to the lake with large wood-made windows and sliding wood panels to darken the spaces inside.

本案是位于意大利波德诺内市。不同层次新街道的出现，以及这里长期被私人占有的事实，迫使部分城区无法开放，因此，遵循将该部分归还给波德诺内市民这一主要议题，拉哥圣乔治城得以重新设计。一条小路将湖泊、公园以及圣乔治教堂广场连接在一起形成公共区域，新的住宅则位于教堂对面。同C+S建筑设计事务所的其他作品一样，本案依然重点关注建筑和景观之间密不可分的关系。建筑地脚线由Prun石材制成，这一材料同样被应用在广场建筑中。

建筑群形态依据湖畔地形变化而变化，所用的材料千变万化：广场由Prun石材制成，湖上小桥采用落叶松木材料，而公园小路则由赭色混凝土铺成。其他公共区域同样体现了建筑师精细的设计理念，门厅和公共楼梯由红色木材制成，礼堂则采用了灰色和红色的隔音板，而教堂的课室则经过特别的平面设计。所有空间均自然采光，并且可以观赏外面迷人的湖光景色。住宅房屋内都安装有大型木质窗户和可滑动木质屏风，既可以很开阔，同时又能与外界空间隔离开来。

Name of Project / 项目名称:
Urban Lake Housing
Location / 地点:
Pordenone, Italy
Area / 占地面积:
Gross Floor Area: 11,000 m²
Site Area: 75,000 m²
Completion Date / 竣工时间:
2010
Project Leader / 项目领导:
Carlo Cappai,
Maria Alessandra Segantini
Design Team / 设计团队:
Carlo Cappai,
Maria Alessandra Segantini,
Carolin Stapenhorst,
Guido Stella
Architecture / 建筑设计:
C+S Associati:
Carlo Cappai,
Maria Alessandra Segantini,
Carolin Stapenhorst,
Guido Stella
Landscape / 景观设计:
C+S Associati
General Contractor / 总承包商:
Rizzani de Eccher S.p.A.
Photography / 摄影:
Alessandra Bello and Luca Casonato
Client / 客户:
San Giorgio S.r.l. (Rizzani de Eccher S.P.A. / Prospettive JV)

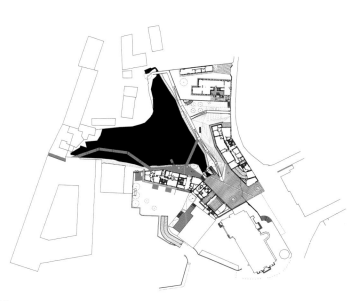

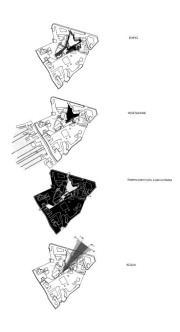

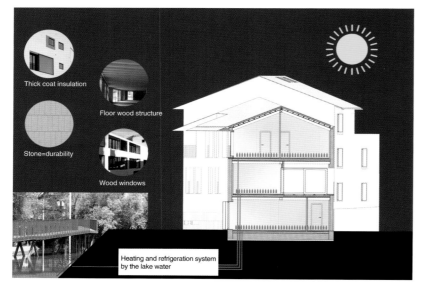

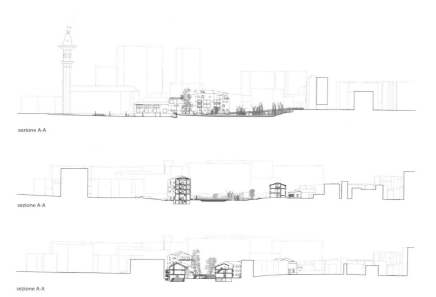

sezione A-A

sezione A-A

sezione A-A

C+S Design has always a very deep attention to sustainability. All materials used in the project are usually specific of the site where the project is found and all buildings are studied for low-carbon emissions.

In this case, the lake itself is not only the expression of the landscape of Pordenone, which is a very beautiful small historical town in the Northeast of Italy, built on the boundaries of the river Noncello, but it becomes also the energetic structure to make the complex work. ULH is designed with a innovative heating and cooling system of pumps exchanging with the different temperature of the lake.

In this sense, the lake itself becomes the main character of the architectural design as it is the background of the new urban scenography, the new liquid soil where to walk over and finally an energetic resource for the buildings' functioning.

长期以来，C+S建筑设计事务所始终关注建筑的可持续性。因此，本案所用材料均取材于当地，所有建筑均进行了低碳排放的研究和试验。

在这种情况下，湖泊本身不仅是波德诺内这个位于意大利东北部Noncello河岸边的美丽小城内的一处景观，它也成为为这个复杂工程中的一个节能元素。一个新型水泵系统的成功设计，可以利用湖水温度的变化，实现制冷和制热。

从某种意义上来看，湖泊本身成为了本案建筑设计中最重要的元素，它不仅是这一新规划区域的背景，方便市民们享受散步时光的地点，同时更是住宅良好运作的能量来源。

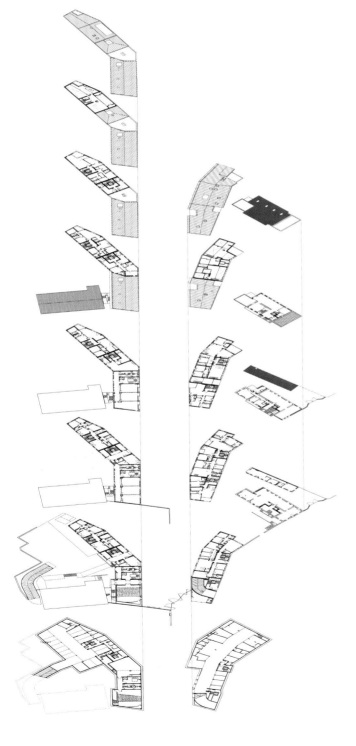

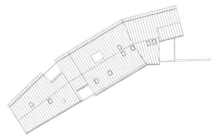

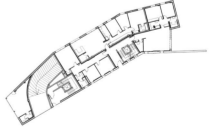
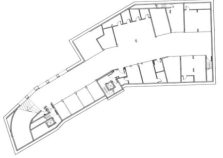

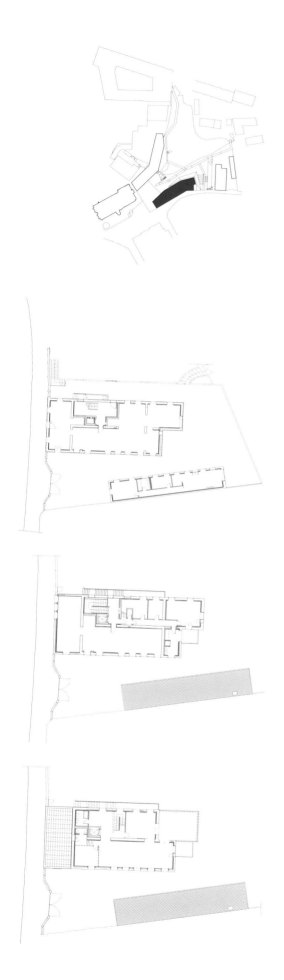

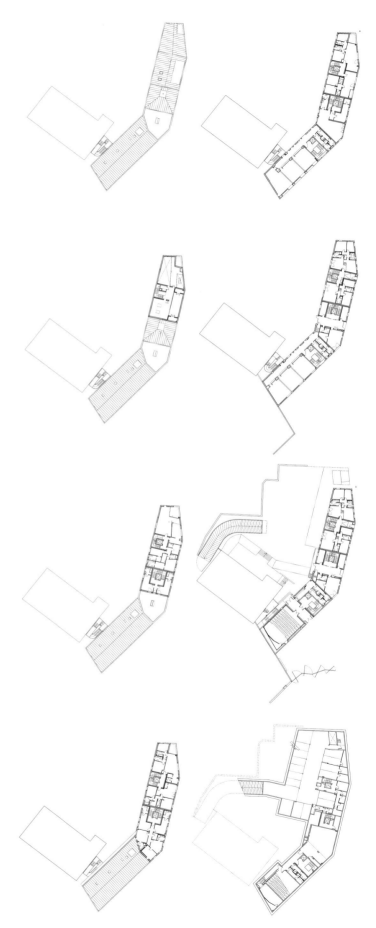

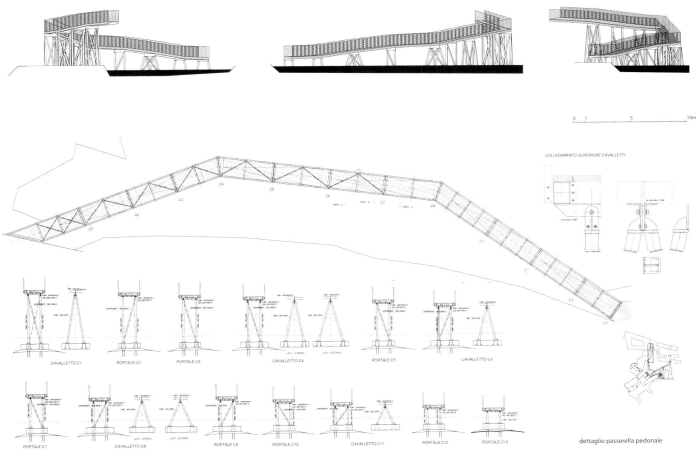

dettaglio passerella pedonale

309

(FER) Studio

Add: 1159 East Hyde Park Inglewood, California 90302 USA
Tel: (310) 672-4749 Fax: 672-4733
Web: www.ferstudio.com

Founded in early 2002, form, environment, research, (FER) studio LLP is an award winning architecture, interiors, landscape, and master planning design firm based in Los Angeles. Noted for its evocative aesthetic and eco-conscious design culture, (FER) studio operates under the direction of Principals, Christopher L. Mercier, AIA, and Douglas V. Pierson, AIA, LEED AP.

(FER) studio's broad range of projects includes academic, civic, retail, restaurant, hospitality and residential, each rooted in the firm's philosophical foundation to reinvent the standards of artistic composition, sustainability, and client collaboration. The art-studio environment fosters a creative design process that blends the arenas of fashion, sculpture, and the relationships between built environments, their users, and their affects on the environment.

Among other projects, (FER) has contributed to Los Angeles landmarks, the Hollywood Roosevelt Hotel, the Thompson Beverly Hills Hotel, Father's Office Culver City, as well as icons of sustainable architecture, such as the Green Building, Louisville's first LEED Platinum pursuant commercial building, and the green Athletics and Performing Arts Centre at the St. Francis School.

ACME

Add: 2-6 Curtain Road, London EC2A 3NQ, UK
T: +44 (0)20 7392 2087
Web: www.acme.ac

ACME was established in 2007 in London. The practice operates in the fields of contemporary architecture, urban planning, interior and product design, it works with private, corporate and public clients.

Anderson Architecture

Add: Simon Anderson Studio 93A Shepherd Street Chippendale NSW 2008 Sydney, Australia
Tel: 02 9319 0224 fax: 02 9690 0908
Web: http://www.andersonarchitecture.com.au

It is an architecture practice based in Sydney Australia that formally begun in 2002 under the direction of Simon Anderson. The firm's inception followed a year living in Stockholm, Sweden which galvanised the aim to marry best practice ecological architecture with contemporary design. A philosophy of humanizing modern architecture and a connection to nature are important in all projects. The firm has won a number of sustainability awards striving to show that good design can also be good for the environment.

Arch. Marcello Albini

Add: Milan, Italy
Email: marcelloalbini@libero.it
Web: www.designrepublic.it/viewdoc.asp?co_id=3208

Since 2005, alternating experiences in collaboration with the architect Emanuela Venturini, Marcello Albini designs furniture and upholstered furniture companies Baxter, Oak Valentini and red lamps and Albizzate Italamp and La Murrina.

He also runs composition of events Swarovski, which has been a consultant for the exhibition in Brussels and Decosit 2006/2007 Index 2006 in Dubai. He designs and supervises the implementation of the interior of homes, shops and showrooms of clothing and footwear in Moscow, in Provence (France), Baku (Azerbaijan), Cyprus and Milan. He produces for Jeckerson (casual wear for men and women and children) the Milan showroom and shop in Moscow. In November 2007 in the Fiera di Rimini dedicated all hotellerie, he designed for the exhibition "Guest Village" and at the same time, designed the space dedicated to duty-free and fashion boutiques and local produce in the airport in Baku. Also in 2007, he joined the Firm Reifig – graphic design in Athens and took up the renovation and furnishing of private homes and shops in Athens and Cyprus.

In recent years, he also taught "Fashion-store design" courses organized by the National Chamber of Fashion and Istituto Marangoni in Milan. In 2008, he still Sia Guest Rimini with the presentation "The Dance of Dreams" between 2008 and 2009, designing furniture, chairs, armchairs and furnishing accessories for Valentino, and Moroso Airnova and even a new lamp for the Murrina the Sea-flower. In 2009, he started a collaboration with Fendi, designing a light chandelier with Swarovski crystals; care coordination MiArt technical event for the evening at Castello Sforzesco in Milan. After the Salone del Mobile, the Three-p will make a new door inserts with Swarovski drowned in the resin.

Architecture International

Add: 225 Miller Avenue, Mill Valley, California 94941, USA
Tel: 415 381 2074; Fax: 415 381 2075
Web: arch-intl.com

Architecture International, Ltd. is a U.S. based architectural design firm providing architecture, planning, and interior design services for clients in the United States, Philippines, China, Italy, India and other parts of the world. The founding Principals, John P. Sheehy, FAIA, RIBA, William J. Higgins, FAIA, and Sherry Caplan, IIDA, Associate AIA, are former Principals of The Architects Collaborative (TAC), a world renown pioneering firm founded by Walter Gropius.

Architecture International, founded in 1994, has adopted the Bauhaus tradition of modernism and multi-disciplinary collaboration. The Principals have established a worldwide reputation for the design of urban mixed-use and high-rise residential design for premier clients. The firm has also been involved in hospitality, housing, retail, business, education and government.

Architecture International's design imagination and creativity allows each project to fully benefit from their all-encompassing experience and working knowledge of sustainability, development, construction technology, materials and costs. This firm not only brings a tenacious spirit of determination to make each project a reality, but they fully take advantage of the specific opportunities of the project, the project goals, objectives, program, and the site. The design and planning process advocated by Architecture International is characterized by extensive, hands-on involvement of the Principals.

Architype

Add: 1b Leathermarket Street, London SE1 3JA, UK
Tel: 020 7403 2889
Web: www.architype.co.uk

Architype has specialised in designing sustainable buildings for over 20 years, with a particular emphasis on the use of natural and healthy materials. Their approach to sustainability is described as "Eco-minimalism" which focuses on "getting the basic priorities right". They adopt a coherent holistic strategy, which makes the form and materials of the building do the majority of the work in reducing energy consumption and of moderating the internal climate, only applying technology when needed.

Arkitektfirma Helen & Hard AS

Add: Arkitektfirma Helen & Hard AS, Vindmollebakken 2,
4014 Stavanger, Norway
Tel: +47 40 64 0672
Email: hha@hha.no
Web: www.hha.no

Helen & Hard was founded in 1996 in Stavanger on the west coast of Norway by architects Siv Helene Stangeland and Reinhard Kropf. Today, the company has a youthful staff of 18 drawn from different countries. The firm works with a wide spectrum of projects from interior, design, art and architecture to town planning. Helen & Hard's work attempts to creatively engage with sustainability, moving away from a solely technical or anthropocentric view towards a synergistic mode, which allows for an extended confluence not only with humans but also with contextual and material resources.

Far from passive ingredients, they have a space-generating capacity as active participants in the design process. This requires a communicative engagement, and synergistic collaboration with particular, local and global resources and knowledge fields. During the design process, they are synthesized in between empirical and experimental sequences, and woven together into architecture.

Helen & Hard's work explores the heterogenic nature of relational space, not as a post-modern narrative collage, but as a fine weave, a field of interrelated sensitive elements, responding to its surrounding environment. This evocative and interwoven space expresses to the user potentiality for interaction and sharpening of all senses.

The endeavor is to create meeting spaces accommodating both, shared social and individual meditative encounter.

Arte Charpentier Architectes

Add: Maison Mozart 8 rue du sentier 75002 Paris, France
Tel: 33 0 1 55 04 1300. Fax: 33 0 1 55 04 1313
Web: www.arte-charpentier.com

Arte Charpentier Architectes is one of the leading and biggest architecture firms in France, with over than 100 professionals worldwide (Paris - Lyon - Shanghai), including architects, urban planners, landscape designers and interior designers.

Founded in 1969 by French architect Mr. Jean-Marie Charpentier, Arte Charpentier Architectes has reached numerous successes and accumulated rich experiences both from theoretical and practical aspects. Its works spread over Europe, Middle East and Asia pacific. Pursuing perfection, innovation, and respecting cultural conversation are the design spirits of the firm.

Charpentier Architecture Design Consulting (Shanghai) Co., Ltd. was set up as a branch office in 2002 in Shanghai to give a better commitment to the rapid development of China. Today, there are over 25 employees in Shanghai led by French and Chinese architects to ensure the originality and the validity of the design work in the office.

BAK Architects

Add: Nuestra Señora del Buen Viaje 1011 1º B, Morón (1708), Buenos Aires, Argentina.
Tel: (+5411) 4-489-5424
Web: www.bakarquitectos.com.ar

BAK Architects is a study integrated by the architects María Victoria Besonias, Guillermo de Almeida and Luciano Kruk. The three of them have experience as architecture's lectures, so each new work results an opportunity to make the entailment between theory and production as a way of overcoming work.

In this way, since the engagement is presented, they begin a searching of solutions joined with reflections based in a few amount of basic principles: they use the common sense to each part of the work, and use simple forms and a few amounts of materials, not only as an aesthetic choice but as an ethical principle of a conscious use of the available materials.

The production of the architecture can not relieve in the apparition of the image, as the physic presence is in charge of give the work sense and emotional character to the spaces. Believe that there are not less important issues in architecture. A simple house or a complex building requires the same enthusiasm to be outlined and demand similar energy to be executed. They confirm in each project that the work of an architect is resolve different problems, and the answer must integrate ethic, aesthetic, spatial and constructive issues.

BDP

Add: 16 Brewhouse Yard, London EC1V 4LJ, UK
Tel: +44 [0]20 7812 8000; Email: enquiries@bdp.com
Web: www.bdp.com

BDP is the largest interdisciplinary practice of architects, designers, engineers and urbanists in Europe.

BDP works closely with users, clients and the community to create special places for living, working, shopping, culture and learning across Europe, Africa, Asia and Australia.

Founded in 1961, BDP now employ more than 1000 architects, designers, engineers, urbanists, sustainability experts, lighting designers and acoustics specialists in studios across the UK, Ireland, Netherlands, UAE, India and China.

BDP has a leading track record in all major sectors including health, education, workplace, retail, urbanism, heritage, housing, transport and leisure. BDP combines expertise across disciplines, locations, sectors and all major building types to deliver a truly integrated way of working – resulting in high quality, effective and inspiring built spaces..

Belzberg Architects

Add: 1507 20th St., Suite C, Santa Monica, CA 90404, USA
Tel: (310) 453-9611; Fax: (310) 453-9166
Web: www.belzbergarchitects.com

Beyond providing exceptional architectural design, there is an inherent understanding that clients entrust their architect to act on their behalf as well as in their best interest, and Belzberg Architects carries out this ideology by integrating the client within the design process to produce an end result which is always more than what is expected. Having not only developed highly individualized designs but realizing said designs through unique and exemplary methods of professional practice is the group ethos. Each project at Belzberg Architects describes their latitudinous way of thinking, which coupled with an intensive look at the contemporary culture of learning and experiencing, defines their approach to the design of large-scale and complex environments. Simply put, they carry out an ideology of integration. Architecture is done at both the local and global scales and the ability to narrate with effectiveness the movement and use of a space brings the individual in and the architecture out.

BIG

Add: Nørrebrogade 66d, 2nd floor,
2200 Copenhagen N, Denmark
Tel: +45 7221 7227; Fax: +45 3512 7227
Web: www.big.dk

BIG is a Copenhagen-based group of architects, designers and thinkers operating within the fields of architecture, urbanism, research and development. BIG has created a reputation for completing buildings that are as programmatically and technically innovative as they are cost and resource conscious. In the architectural production they demonstrate a high sensitivity to the particular demands of site context and programme.

As designers of the built environment, the group test the effects of scale and the balance of programmatic mixtures on the social, economical and ecological outcome of a given site. Like a form of programmatic alchemy they create architecture by mixing conventional ingredients such as living, leisure, recreation, working, parking and shopping to realize imaginative and responsible solutions. Their philosophy recognizes the added value to be brought to each building site and the recipe of programmatic alchemy allows for the development of constructible

Bohlin Cywinski Jackson

Add: Wilkes-Barre, Pittsburgh, Philadelphia,
Seattle, San Francisco, USA
web: www.bcj.com

Bohlin Cywinski Jackson, founded in 1965, has offices in Wilkes-Barre, Pittsburgh, Philadelphia, Seattle and San Francisco. The firm's work is known for exceptional design, for its commitment to the particularity of place and user, and for an extraordinary aesthetic based on a quiet rigor which is both intellectual and intuitive.

The firm's work ranges greatly in scale and circumstance. Its architecture is alive to the subtleties of place — man-made or natural, to the varied natures of people, to the sensibilities of individuals, to the character of institutions, and to the rich possibilities of materials and the means of construction.

Bohlin Cywinski Jackson has received more than 460 regional, national and international awards for design. In 1994, the practice received the Architecture Firm Award from the American Institute of Architects. The firm's work is published regularly in professional journals worldwide.

C+S Associati

Add: Piazza San Leonardo, 15, 31100 – Treviso – Italy
Tel: +39.0422.591796, Fax: +39.0422.591796
Email: press@cipiuesse.it
www.cipiuesse.it

Carlo Cappai and Maria Alessandra Segantini founded their architecture office C+S ASSOCIATI in Treviso, Italy, in 1994. They have been visiting professor in architectural design in several universities and they won many international awards. Their work is published in the most important architectural reviews such as: 2G (E), Abitare (I), AD (GB), Area (I), A+U (JA), Architectural Review (GB), AW (DE), Bauwelt (DE), Casabella (I), D'Architettura (I), Detail (DE), Domus (I), L'Architecture d'Aujord'hui (F), Loggia (ES), Spazio e Società (I) and The Phaidon Atlas of 21st Century World Architecture (UK). Their work was exibited in Venice (8th and 12th Biennale of Architecture and 50th Art Biennale), Milan, Paris, London, Essen, Vienna, Piran, Muenchen.

Calder Flower Architects

Add: Level 2 140 Myrtle Street, Chippendale NSW 2008 Australia
Tel: +61 2 9698 9822; Fax: +61 2 9698 5937
Web: www.calderflower.com.au

Calder Flower Architects is a Sydney-based practice with thirty years experience in a range of community architecture including houses, multi-residential and aged-care developments. The practice is expert and active at all levels of the design delivery process from interrogation of the brief to construction administration. CFA is committed to promoting environmentally responsible and environmentally responsive design. The practice uses a variety of media to conceive, develop and test design work including sophisticated 3D computer modelling and gaming technologies. As well as a consistent inventory of current building projects, regular participation in theoretical design competitions allows the practice to continually test and develop new ideas.

Christina Zerva Architects

Add: 383 N. Glebe Rd, Arlington, VA 22203, USA
Tel: +1 202 355 5229
Web: www.christinazerva.com

Christina Zerva Architects, founded in 1991, is a worldwide practice, working globally with project offices in Washington DC, Larissa and Belgrade. Christina Zerva leads the office with partners Paola Nena Nastasia Sakic and Milan Li. The practice works internationally on residential, cultural and commercial projects. With each project they explore new ways to ensure that the best results are achieved through integration of concept and context with the programmatic and functional essence of a building. Through the continuity of the design process and extensive research involved in the analysis of site, program, social context, emerging construction and contemporary technologies, the approach and methodology introduce new perspectives and new dynamics, reinvigorating both landscape and cityscape.

They provide full architecture, master planning, project management, interior design, product and furniture design services for both the private and public sectors. They use materials and sense of order that constitute the discipline and practice of modern architecture. The practice has initiated systematic research into a responsible approach on systems that reduce energy consumption and regulate harmful emissions. Their architecture seeks to reduce waste of energy, water and materials. By taking advantage of a system of holistic design using technologies, methods and products that not only protect your health and the world around you, but actually enhance the spiritual, emotional, and mental quality of your daily experience.

Coates Design, Inc.

Add: PO Box 11654 Bainbridge Island, WA, USA
Tel: 206 780 0876
Web: http://www.coatesdesign.com

Coates Design, Inc. is gaining a sterling reputation for offering innovative architectural and sustainable design services for a wide range of projects. Coates pride themselves on consistently exceeding their client's goals, on both large and small construction projects. Their highly creative, qualified team members, with backgrounds in design, construction and the fine arts, transform complex design challenges into elegant solutions.

Award-winning Architect, Matthew Coates, AIA, is recognized by the U.S. Green Building Council (USGBC) as a LEED accredited professional. Let Coates Design, Inc. create sustainable and environmentally sensitive solutions for your specific design or building needs.

Dattner Architects

Add: 1385 Broadway, 15th Floor, New York,
NY 10018, USA
Email: Info@dattner.com
Web: www.dattner.com

Dattner Architects is a New York City firm founded in 1964. The firm's work includes master planning and buildings for educational and cultural institutions, public agencies, not-for-profit groups and corporate clients, historic preservation/adaptive reuse, sustainable architecture, interior design, and spans from small scale interiors to large urban planning projects.

With each project, regardless of size or type, they aim to realize their clients' highest aspirations while respecting shared social responsibility and building within available resources. Since its inception, the firm has been characterized by the continuing involvement of the firm principals, from the beginning of the project through post-occupancy services.

David Hertz Studio EA

Add: 1920 Olympic Blvd Santa Monica, CA 90404, USA
Tel 310.829.9932; Fax 310.829.5641
Web: http://www.studioea.com

The work of David Hertz Architects Inc, Studio of Environmental Architecture focuses on the design and construction of environmentally responsible residential and commercial buildings, and engages in multiple facets of design, including product design, furniture design, public sculpture, environmental consulting, as well as the research and development of resource efficient ecologic building products.

In 1983, Hertz founded Syndesis Inc, and invented the material manufacturing process of Syndecrete®, the innovative pre-cast lightweight concrete surfacing material. Hertz sold the technology in 2006. Evolving from a generalist nature of practice to a more specified studio has allowed for a comprehensive understanding of creative problem solving and the opportunity to view design challenges from multiple perspectives simultaneously.

S.E.A. has been located in Santa Monica for over the last 20 years. It is a small practice dedicated to design and green building, and employs about 10 people.

Dick Clark Architecture

Add: 207 West 4th St Austin, TX 7870, USA
Tel: 512.472.4980; Fax: 512.472.4991
Web: www.dcarch.com

Dick Clark Architecture was founded in 1979. For over 30 years, the office has been influenced by the creative environment and technology that have driven Austin. Dick Clark Architecture was formed here, but has grown parallel with the city to expand their influence and work. Creative and engaged in the community, the staff continues to draw inspiration from Austin for work that contributes to the shaping of central Texas and beyond. Local knowledge, creativity and an ability to work openly and collaboratively produce remarkable results.

Their work covers a wide range of project types from master planning and site feasibility studies to custom single-family homes. Their commercial work includes offices, hotels, mixed-use projects, restaurants, multi-family housing and retail developments. For the commercial work, they start with the understanding of the specific business goals for each owner and apply their creative process to help them to surpass their expectations. Dick Clark Architecture's custom residential owners appreciate their collaborative process to design a home that expresses each family's unique and individual qualities.

Fernanda Marques Associated Architects

Add: Rua Ramos Batista, 198 – Cobertura,
São Paulo 04552-020, Brazil
Tel: +55 11 3849 3000; Fax: +55 11 3848 3493
Web: www.fernandamarques.com.br

Fernanda Marques graduated in Architecture and Urbanism, at the University of São Paulo, in the year of 1988. She also made specialization courses in US and Europe.

Founded in 1990, Fernanda Marques Associated Architects is internationally recognized by the contemporary and minimalistic style of its architecture, features also seen on her interior design projects. Fernanda´s Architecture Office is installed in two floors of a modern building in Vila Olímpia, São Paulo – Brasil. The office has a team with more than 70 professionals and is sectorized in order to attend all the architecture´s market demands. During 2009, the office realized more than 180 projects, including 80 real estate developments, 50 residential, 30 corporate and 20 architecture events.

Francis-Jones Morehen Thorp

Add: Level 5, 140 George Street,
Sydney NSW 2000, Australia
Tel: +612 9251 7077; Fax: +612 9251 7072
Email: media@fjmt.com.au
Web: http://www.fjmt.com.au

Francis-Jones Morehen Thorp is a multi-award-winning Australian architectural practice dedicated to design excellence and the enhancement of the public domain. The multi-disciplinary fjmt design studio explores the evolution of architectural form by synthesising place and programme through an elaboration of the tectonic. fjmt's work focuses on the spatial and organic interconnection of built form and site to embody shared values and aspirations.

fjmt have won numerous international and national architectural competitions and design awards including the AIA Sir Zelman Cowen Award for Public Buildings, the Lloyd Rees Award for Urban Design, the Lachlan Macquarie Award for Heritage, the National Award for Sustainable Architecture and runner-up for the Emilio Ambasz Award for Green Architecture.

Through studios in Sydney, Melbourne and Oxford, and project offices interstate and overseas, fjmt undertake public, institutional, commercial and residential commissions throughout Australasia and Europe. These commissions are frequently the result of international design competitions.

FTL Design Engineering Studio

Add: 44 East 32nd Street 3rd Floor New York, N.Y. 10016 USA
Tel: 212 732 4691
Web: http://www.ftlstudio.com

As FTL Design and Engineering Studio's principal, Goldsmith has designed and led the firm's legacy projects including portable and permanent buildings such as the Carlos Moseley Pavilion for the Metropolitan Opera and The New York Philharmonic, the DKNY World Headquarters, and the Cirque du Soleil Structures for Disney World.

In recent years, Goldsmith has expanded the firm's focus from tensile architecture to applying his design philosophy and approach to new materials such as polycarbonate cladding, steel cable nets and foil pillow structures. Recent projects include, SkySong, a 4645 square metres shade structure for a mixed-use office, research, retail and hotel conference complex in Scottsdale Arizona's Innovation Centre, the Rosa Parks Transit Centre in Detroit, Michigan, and the new Sun Valley Pavilion, a performance amphitheater in Idaho. His design for the Santa Fe New Mexico Opera's Plaza structure recently received the 2008 Industrial Fabric International Outstanding Achievement Award.

Fumie Shibata, Creative Director

Add: 4-2-35-301, Roppongi, Minato-ku, Tokyo, 106-0032, Japan
Tel: (81)3 3479 7113; Fax: (81)3 3479 7103
Web: http://www.design-ss.com

Fumie Shibata is a Product Designer based in Tokyo. Focusing on product design, she has expanded her design activities in industrial design from daily goods to electronics and healthcare products. Her works have received tremendous acclaim in various design awards worldwide including, Gold Prize of if Award, red dot Design Award in Germany and the Good Design Award in Japan. 9h, nine hours owes its design style to Fumie Shibata, General Creative Director.

The duo that made her design concept a reality were Masaaki Hiromura as Graphic and Sign Designer and Takaaki Nakamura as Interior Design. To give her design cohesiveness, she also creates all 9h's original product line.

GAD

Add: Tesvikiye Cad, 3B Gunes Apt, Tesvikiye Istanbul 34367 Turkey
Tel: +90 212 327 5125; Fax: +90 212 258 1663
Web: www.gadarchitecture.com

Global Architectural Development is an Istanbul and New York based company, which performs architectural practice, research and concept design since 1994 and whose partners are Gokhan Avcioglu and Ozlem Ercil Avcioglu and their global collaborators.

Contemporary and current architecture, urbanism, software, consumer habits and behaviors and approaching to the projects holistically are among its field of interest. GAD understands architecture as a practice that relies on the experiment, values historical precedents and new ways to combine both in a mutually benefiting fashion.

GAD Architecture is committed to finding innovative approaches to architecture and creating new spatial experiences with projects and ideas. GAD has won numerous awards including the 1997 Turkish Architecture Prize for the design of a Public Park in Istanbul and the 2001 Cimsa Design Prize for outdoor seating, a bronze medal in Miami Biennale for Borusan Exhibition Centre in 2003. The Public park in 1998 and in 2004 and Esma Sultan Venue place in 2001 was short listed for Aga Khan Award for Architecture. GAD has offices in Istanbul, New York and Bodrum.

Gonzalo Mardones Viviani

Add: Av. del Valle 869 of 01 Huechuraba, Santiago, Chile, 8580000
Tel: +56 2 9493081 Fax: +56 92 9493086
Email: gmf@gonzalomardonesv.cl
web: www.gonzalomardonesv.cl

Gonzalo Mardones Viviani was born in Santiago de Chile on July 8th 1955. He gets his degree as architect from the Universidad Católica de Chile, where he graduates with the Maximum Honors. He receives the First Prize in the Architecture Biennale, for the best degree project among all the Architectural Schools in Chile, for his project for urban renewal of the South-West Centre of Santiago.

He has been a professor of architectural design workshops and directed degree projects in the Faculties of Architecture of the Universidad Católica, Universidad de Chile, Universidad Central, Universidad Andrés Bello and Universidad Finis Terrae, in addition to having been guest professor and lecturer in different universities in Chile, and abroad.

His work has been published by the main architectural magazines and honored at Biennales. He has been a member of the National Commission of Competitions of the Architects Association in Chile and a Founding member of the Association of Architectural Practices (AOA).

Hays + Ewing Design Studio

Address: 609 East Market St., Suite 203 Charlottesville, VA 22902 USA
Tel: 434-979-3222
Web: www.hays-ewing.com

Hays + Ewing Design Studio is an award-winning architecture firm that merges inspiring design with green building technologies in commercial, cultural, residential, planning and institutional projects. Based in Charlottesville, Virginia, the firm seeks to fulfill client's needs and desires through a highly collaborative approach. Christopher Hays and Allison Ewing formed Hays+Ewing Design Studio in 2004 and focus on the careful integration of buildings with their environment. Before starting their own firm, they worked with internationally-recognized firms Cesar Pelli & Associates, Renzo Piano Building Workshop and William McDonough + Partners.

At these firms and as partners at William McDonough + Partners, they worked on a variety of project scales and types in projects located around the globe. This experience has invested them with the ability to tackle a diverse range of projects, climates and sites with a unique vision specific to each – one that is at once innovative yet pragmatic.

Hays and Ewing, both LEED accredited professionals, have worked in the field of sustainable design for over eighteen years, as former partners of the firm of William McDonough + Partners, where they led numerous LEED certified projects, and as principals of their own firm. Their focus on the environment operates at several levels -- in harmonizing buildings with their surroundings, and in the integration of building and natural systems, such as cooling through natural ventilation, rainwater collection, and harnessing energy from the sun.

Héctor Ruiz-Velázquez Architecture

Add: C/ Fernando el catolico 6, 6- 2 dch Madrid Spain 28015
Tel: 34 91 5774518
Web: www.ggrvarquitectos.com

Hector Ruiz-Velazquez has a degree in architecture from the University of Virginia, USA. With studio professors from Harvard University and Columbia University, Hector Ruiz-Velazquez has founded his own architectural office in 1992 as a culmination of an extensive professional practice that includes architectural projects of big scale to corporate image.

A characteristic of his studio is its diversity not only cultural but also regarding the professional disciplines. The studio covers projects from urbanism to graphic design, industrial and interior design, photography, as well as integral corporate images.

Hector Ruiz Velazquez' projects have been published in numerous publications all over the world since 1992 and he has been invited to lecture in different universities and public and private institution in different parts of the world.

IAAC

Add: C/ Pujades, 102, Barcelona, Spain
Tel: +34 93 320 9520
Web: www.iaac.net

The Institute for Advanced Architecture of Catalonia (IAAC) is an international centre for research, education, investigation and development, oriented toward architecture as a discipline that addresses different scales of territorial analysis and urban development as well as diverse architectural projects, digital processes and information environments.

Located in Barcelona, one of the international capitals of Ur¬banism, the institute directed by Vicente Guallart develops multidisciplinary programmes that explore international urban and territorial phenomena, with a special emphasis on the op¬portunities that arise from the emergent territories and on the cultural, economic and in social values that architecture can contribute to society.

IAAC sets out to take R+D to architecture and urbanism and to create multidisciplinary knowledge networks, and to this end the institute works in collaboration with a number of cities and regions, industrial groups and research centres, including the Massachusetts Institute of Technology (MIT), the University of Brighton and the University of Chicago, developing various research programmes which bring together experts in different disciplines such as engineering, sociology, anthropology, architecture and other fields of investigation. IAAC has made a name as a centre of international reference (as shown above) which this year welcomes students and investigators from 24 countries, among which are India, Australia, the USA, Poland, Argentina and Iraq.

The IAAC is housed in an old factory building, with 2,000 square meter of space for research, production and dissemination of architecture, so that the space itself is a declaration of principles, embodying an experimental and productive approach to architecture. The IAAC is engaged in a variety of research projects as well as workshops and courses, and special summer workshops, open to Spanish and international firms and institutions.

Ignatov Architects

Add: New York, USA / Varna, Bulgaria
Tel: +1 212 685 2620 USA / +359 884 335 810 Bulgaria
email: studio@bignatov.com
Web: www.bignatov.com

"Seeing far is one thing, going there is another" Constantin Brancusi. Ignatov Architects is a mobile architectural practice interested in finding inventive and efficient solutions. The firm provides complete architectural services from site analysis to design and construction administration. Ignatov Architects believe that meaningful architecture must be well-thought and specific to its purpose, location and users.

The company was established by Mr. Boris Ignatov in 2004 in New York. Mr. Ignatov holds Master degrees in architecture from Columbia University in New York (2006) and Sofia University of Architecture (1995). He has worked for several large architectural firms including "Perkins + Will Architects" and is a licensed architect in the state of New York and Bulgaria

Jakob + Macfarlane Architects

Add: 13, rue des petites écuries 75 010 Paris, France
Tel: 33 (0) 1 44 79 05 72, Fax: 33 (0) 1 48 00 97 93
Email: info@jakobmacfarlane.com
Web: www.jakobmacfarlane.com

Jakob + Macfarlane, Architects, urbanists, designers was cofounded by Dominique Jakob and Brendan Macfarlane. Dominique Jakob: In 1990, Dominique Jakob received her degree in Art histoiry from Université de Paris I. In the later year, she received her degree in Architecture from the Ecole d'Architecture Paris-Villemin. From 1998 to 1999, she was a visiting teacher at Ecole spéciale d'Architecture de Paris and from 1994 to 2004, she also tought in Ecole d'Architecture de Paris-Villemin / Malaquais. Brendan Macfarlane: In 1984, he achieved Bachelor of Architecture from Sci-Arc (South Calif. Inst. of Architecture). In 1990, he graduated from Graduate School of Design of Harvard University. From 1996 to 2008, he gave lectures in many colleges around world including Graduate School of Design, le Berlage Institute, Bartlett School of Architecture, l'Ecole Spéciale d'Architecture, Berlage Institute, Sci-Arc à Los Angeles, Harvard Design Graduate School and University of Florida.

JGA

Add: 29110 Inkster Road, Suite 200 Southfield, MI 48034 USA
Tel: 248.355.0890 ; Fax: 248.355.0895
Web: www.jga.com

JGA has evolved to become a leading global retail design, brand strategy and architectural firms. Since 1971, JGA has built its reputation by helping retailers realize their visual marketing potential and attain leadership within their niche. JGA believes that bringing a creative idea into reality and achieving success requires the integration of strategic clarity, competitive and market awareness, conceptual innovation and a strong business sense. As strategists and designers JGA's services are extensive. A team approach offers significant value over conventional firms.

As part of the client's project team, JGA motivates the creative process of provoking ideas and solutions and facilitating the management process through strategic design, scheduling, budgets and implementation, resulting in increased consumer interaction and satisfaction. JGA offers a diverse menu of services including market and design strategy, conceptual positioning, visual communication design and logo/brand identity, design and architectural development and implementation, retail tenant coordination, construction administration and fixturing/furniture and materials procurement.

Jorge Hernandez de La Garza

Tel: 52 55 5211 0045

Graduated from Universidad La Salle de Mexico. In 1999, he won the Architectonic Composition Award Ing. Alberto J. Pani. In 2002 Studied Design at AA The Architectural Association in London, England. In 2005 is finalist of the Icons of Design Awards with Los Amates house. In autumn of 2006 wins the Icons of Design Awards with the project of Vladimir Kaspe Cultural Centre, and also this year is finalist of the Interior Design Award with the project Showroom Comex. In 2007 was selected as one of the 44 international firms for the Young Architects Annual Event in Spain. In this same year was part of the 101 Most Exciting New Architects In London. In 2008, the College of Architecture of Mexico City gives him the first mention of Young Architects under 40 years old. His works have been published in Tokyo, England, Spain, Portugal, Brazil, Korea, Ukraine, Russia, Argentina, Italy, Germany and México.

Judd Lysenko Marshall Architects

Add: 7 Glenard Drive Eaglemont VIC 3084,Melbourne, Australia
Tel: 0411 214 832
Email: j.judd@jlma.com.au
Web: www.jlma.com.au

Judd Lysenko Marshall is a new and energetic design consultancy committed to innovative and imaginative solutions in the fields of architecture and urban design. Their diversely skilled team brings creative thought, scholarship and intelligence to each project. Judd Lysenko Marshall is focused on the delivery of inspired design responses grounded in an absolutely functional architectural approach. They strive for maximum impact within tight budgets. They aim to reveal a complete understanding of the diverse and specific needs of each and every client. Through creative thought, use of new technology and just plain hard work, they produce projects of environmental sustainability, true excitement and life long durability.

Karim Rashid

Add: 357 West 17th St, New-York, 10011, USA
Tel: 212.929.8657; Fax: 212.929.0247
Web: www.karimrashid.com

Karim Rashid is one of the most prolific designers of his generation. Over 3000 designs in production, over 300 awards and working in over 35 countries attest to Karim's legend of design. His award winning designs include democratic objects such as the ubiquitous Garbo waste can and Oh Chair for Umbra, interiors such as the Morimoto restaurant, Philadelphia and Semiramis hotel, Athens and exhibitions for Deutsche Bank and Audi. Karim has collaborated with clients to create democratic design for Method and Dirt Devil, furniture for Artemide and Magis, brand identity for Citibank and Hyundai, high tech products for LaCie and Samsung, and luxury goods for Veuve Clicquot and Swarovski, to name a few. Karim's work is featured in 20 permanent collections and he exhibits art in galleries world wide. Karim is a perennial winner of the Red Dot award, Chicago Athenaeum Good Design award, I.D. Magazine Annual Design Review, IDSA Industrial Design Excellence award.

Kevin Roche John Dinkeloo and Associates LLC

Add: 20 Davis Street / P.O. Box 6127,
Hamden, CT 06517-0127, USA
Tel: 203-777-7251; Fax: 203-776-2299
Web: www.krjda.com

The firm of Kevin Roche John Dinkeloo and Associates LLC (KRJDA), located just outside New Haven, Connecticut, is a direct outgrowth of Eero Saarinen and Associates, which was originally established in 1950. When Eero Saarinen died in September 1961, the practice was taken over by Kevin Roche and John Dinkeloo with Kevin Roche resolving the remaining design issues of the twelve major projects on which Mr. Saarinen had been working at the time of his death. These included the Dulles International Airport, the St. Louis Gateway Arch, and the CBS Headquarters in New York.

In 1966, the firm became Kevin Roche John Dinkeloo and Associates. John Dinkeloo died in June of 1981, and Mr. Roche continues the practice with the original firm name. The firm now consists of three principals: Kevin Roche, FAIA, Philip Kinsella, AIA and James Owens, AIA. In addition, there is a total staff of 50. Kevin Roche John Dinkeloo and Associates is engaged in major projects throughout the United States, Europe and Asia and provides complete master planning, programming, architectural design, interior design, working drawings, specification and construction administration services.

Kevin Roche, who is responsible for all of the designs for the firm, has designed a variety of institutional and corporate projects including 38 corporate headquarters, three hotel/apartment buildings, eight museums, numerous research facilities, theaters, schools, factories, performing arts centers, houses and the Central Park Zoo in New York. For the past 44 years, he has been the architect for the master plan and expansion of the Metropolitan Museum of Art in New York, designing all of its new wings and installing many of its collections.The firm received the American Institute of Architects 1974 Architectural Firm Award and more recently the firm was the recipient of the 1995 American Institute of Architect's 25-Year Award for the Ford Foundation Headquarters in New York City. In 1982, Kevin Roche received the Pritzker Architectural Prize and in 1993, he received the AIA Gold Medal.

Kjellgren Kaminsky Architecture

Add: Ekmansgatan 3, 411 32 Göteborg, Sweden
Tel: +46 (0) 31 761 20 01; Fax: +46 (0) 31 18 21 04
Web: www.kjellgrenkaminsky.se

Kjellgren Kaminsky Architecture is an award winning architecture firm based in Göteborg, Sweden. They work with architecture in its broadest meaning ranging from furniture to city planning, from theory to practice.

Every project has its own story. It starts with a demand for change within a context. From this point of departure they narrate the process in collaboration with the client. In the end they aim to create functional and sustainable architecture with a poetic dimension.

Kjellgren Kaminsky Architecture is a network based company. For every project competences from different fields are fused. The core is made up of the two founders and a total staff of eight architects and engineers and their skills in architecture and interior design. The network also extends outside the company where they collaborate with specialists in the fields of sustainable development, visualization and engineering. This organization renders a highly flexible and creative environment.

Laetitia Delubac and Christian Félix Architects

Add : FELIX-DELUBAC architectes, 7, rue Moncey,
75 009 Paris, France
Tel: +33 (0)1 49 70 04 62
E-mail : contact@felix-delubac-architectes.com
Web: www.felix-delubac-architectes.com

Laetitia Delubac and Christian Félix architects was founded byLaetitia Delubac and Christian Félix. They started working together on open architectural competitions as early as the architecture school.

Activities : R&D, about designing accommodations for individuals (emergency housing, ecological housing, micro- architecture) and designing original furniture. At present they are designing a project of a passive house built with wood and thatch for Normandy.

LAN Architecture

Add : 25 Rue d'Hauteville 75010 Paris, France
Tel: +33 1 43 70 00 60 ; Fax : +33 1 43 70 01 21
Web: www.lan-paris.com

LAN (Local Architecture Network) was created by Benoit Jallon and Umberto Napolitano in 2002, with the idea of exploring architecture as an area of activity at the intersection of several disciplines. This attitude has developed into a methodology enabling LAN to explore new territories and forge a vision encompassing social, urban, functional and formal questions. LAN's projects seek to find elegant, contemporary answers to creative and pragmatic concerns.

LAN has received several awards: the Nouveaux Albums de la Jeune Architecture (NAJA) prize awarded by the French Ministry of Culture and Communication (2004); the International Architecture Award from the Chicago Athenaeum and the European Urban Centre for Architecture, Art, Design and Urban Studies, the Archi-Bau Award, the Special Prize at the 12th World Triennale of Architecture, Sofia (2009); the AR Mipim Future Projects Award and the Europe 40 Under 40 Award (2010).

LAVA

Add: 72 Campbell Street, Surry Hills,
Sydney, Australia
Tel: +61 2 92801475; Fax: +61 2 92818125
Web: www.l-a-v-a.net

The Laboratory for Visionary Architecture [LAVA] was founded by Chris Bosse, Tobias Wallisser and Alexander Rieck in 2007. LAVA breaks new ground by exploring frontiers that combine future technology with organic and natural structures and patterns of organisation. LAVA combines digital workflow, contemporary materials and the latest digital fabrication technologies with the aim of achieving "more with less": more (architecture) with less (material/ energy/time/cost). Merging new technologies with the principles found in nature will result in a smarter, friendlier, more socially and environmentally responsible future. LAVA recently won a UN partnered ZEROprize Award for the reskinning of sixties buildings and an international competition to design the centre of the world's first zero carbon city.

Lean Arch, Inc. Design + Construction

Add: 155 West Washington Blvd., Suite 1202, Los Angeles,
CA 90015 USA
Tel: 213.744.9830, Fax: 213.744.0123
Web: www.leanarch.com

The architectural design/build practice of Lean Arch, Inc. was established in the summer of 2000 and officially incorporated in August 2001. Founder James Meyer describes the office as a design/build collaborative focusing mostly on residential, commercial and civic-oriented architecture, but also as having worked on a variety of projects ranging from furniture and product design to corporate branding, packaging and graphics. Founded on the principles of modernism, the firm strongly believes that architecture and the built environment have the ability to improve people's lives – their physical surroundings, their appreciation of nature, their understanding of the arts, and their relationships with one another. The firm is committed to offering ideas that address the specific needs of a client while remaining conscious of the physical and psychological impact a work will have on both nature and the human-made environment.

Lean Arch is a full-service firm offering architectural design and construction services. The office prides itself on the quality and extent of service provided to clients. All projects, regardless of scale, are given the same level of attention – "right down to the doorknobs."A variety of techniques are used in the presentation of design ideas. These may range from sketches on paper to more realistic computer-generated 3D models and animations. From the conceptual or schematic phase through construction, Lean Arch works closely with clients, consultants, and contractors to make sure that what has been designed is what will be constructed. Objectives have always been met by keeping in mind the idea that people must work together to solve a problem. More ideas, better solutions…

Lehrer + Gangi Design + Build

Add: 2140 Hyperion Ave- Los Angeles, CA 90027-4708, USA
Tel: 323.664.4747, Tel: 323.664.3566
Web: www.lehrerarchitects.com

Lehrer + Gangi Design + Build was created by Michael B. Lehrer, FAIA, Mark Gangi, AIA, and Frank P. Gangi, MBA, for the Diamond Valley Lake museums project. The collaboration between the two architects and builder, forming a design/build alliance, was instrumental in bringing this project in on time and on budget. Lehrer + Gangi Design + Build was hands-on from initial conception, through design, building, and LEED commissioning. Lehrer, principal of Los Angeles-based Lehrer Architects, is well known for his community-based work, such as the award-winning James M. Wood Community Centre and Downtown Drop-In Centre, both in Downtown LA. Gangi has been principal of Burbank-based Gangi Architects since 1990, and has produced an array of urban projects.

Line and Space

Add: 627 E. Speedway Blvd., Tucson, AZ 85705 USA
Tel: (520) 623-1313
Web: www.lineandspace.com

Line and Space, LLC was founded in 1978 in Tucson, Arizona by Les Wallach, FAIA, to facilitate the design and building of innovative and ecologically sound architectural projects. The firm strives for the quiet integration of structure and landscape - projects encourage and demonstrate notions of environmental stewardship. Being students of regional cultural influences, they avoid stereotypical "fantasy" responses in favor of the deeper meaning of metaphor, entry sequence, scale, materials, colour and form; all of which are important in successful architecture. For over thirty years, Line and Space has studied and perfected their understanding of the climate and the land. Extensive programming helps to define the true needs of both the client and the site. Their architecture shows reverence to the textures and form of the fragile environment. Their engineering recognizes the cooling effects of shade, shadow, and the efficient use of water and wind in harsh surroundings. Their buildings take advantage of native materials and their affordability. The power of stone, masonry and stucco, sensitively juxtaposed with the delicacy of wood, metal and colour, are a tribute of their mastery at building. Such indigenous materials play an important role in the organic relationship that welds structure to the earth; how buildings touch the sky contributes to their visual success. Above all, the work of Line and Space demonstrates that architecture is an art – one that should please the senses while uplifting the spirit.

Marcus O'Reilly Architects

Add: 19 Baker St, St Kilda VIC 3182 Australia
Tel: +61 3 9534 3715, Fax: +61 3 9537 3715
Web: http://www.marcusoreilly.com

Marcus O'Reilly Architects is a small award winning practice with a strong design focus. The practice provides a high level of personal service and engagement and has completed work that ranges from new homes and renovations to commercial buildings and public commissions.

Marcus O'Reilly Architects work with clients to develop their brief. With the client's ideas as a starting point, Marcus O'Reilly Architects then applies their creative and practical expertise to create a fresh, tailor-made approach. Their unique projects are deeply rooted in their context and are tied to the strongly held belief that the built world should reflect a sense of place.

Their work demonstrates a strong commitment to good design. In addition, their practice offers the following expertise: Thorough knowledge of the planning approval process, including its vagaries. Realistic costing and cost-controlling strategies, ensuring projects come within their targeted budgets. Sound understanding of building techniques and materials. Interest in expanding knowledge and application of innovative techniques, particularly in the area of sustainable design. Tight documentation. Effective contract administration.

Max Pritchard Architect

Add: Max Pritchard Architect PO Box 808
Glenelg SA 5045 Australia
T: 61 8 8376 2314 Fax: 61 8 8376 2317
Web: http://www.maxpritchardarchitect.com.au

The works of Max Pritchard speaks eloquently of its unassuming creator: deft, grounded and genuine. It is appropriate in a time when the profession is going through a period of self-examination of its role and relevance in South Australia that they recognize a person who has quietly built a reputation and following as an architect of people and place, unaffected by the transient trends and fashions swirling around him.

Max Pritchard's work has been consistently reconised by his peers, having received many RAIA Merit and Commendation awards, and by interstate and overseas architecture critics and commentators.

Moho Arquitectos

Add: Paseo Fotografo Verdú 8 Bajo D,
30002 Murcia, Spain
Tel: +34 868 550 316; Fax: +968 210 700
Web: www.mohoweb.com, www.cartonlab.com

Joy is the engine of growth and the inspiration of the work. Laugh is a measure of how comfortably they express themselves. They like to think about their projects as beautiful experiments, iterations, attempts, trials, and errors. Little jumps over irrational tradition and disciplinary fences.

Mork-Ulnes Design

Add: Mork-Ulnes Design (MU/D), 602 Minnesota Street,
San Francisco CA 94107, USA
Tel: 415-282-1437; Fax: 415-206-9342
Web: www.mork-ulnesdesign.com

MU/D is a multi-disciplinary based design firm led by Casper Mork-Ulnes. Norwegian born, Casper was raised in Norway,

Italy, and the United Kingdom. The firm's design philosophy is rooted in this international background and is reflected in its geographically varied portfolio. A profound interest in place and the peculiarities of each site, combined with ingenious use of materials and specific program requirements generate varied and unique designs. In addition to his responsibilities at MU/D, Casper is also a Founder and Partner in the Pre-Fab building company Modern Cabana, also in San Francisco.

He holds a Masters degree in Architectural Design from Columbia University and a Bachelor of Architecture degree from California College of Art. He is a licensed architect in the European Union and has built projects on three continents. In his over 15 years of design work, Casper has worked on projects for clients including: BMW, Swatch, Norwegian Ministry of Foreign Affairs, Dominican University, University of the Pacific/McGeorge School of Law and private developers.

MVRDV

Add: Dunantstraat 10, PO Box 63136,
NL - 3002 JC Rotterdam, the Netherlands
Tel: +31 (0)104772860; Fax: +31 (0)104773627
Web: www.mvrdv.nl

MVRDV was set up in Rotterdam (Netherlands) in 1993 by Winy Maas, Jacob van Rijs and Nathalie de Vries. In close collaboration the 3 principal architect directors produce designs and studies in the fields of architecture, urbanism and landscape design. Early projects such as the headquarters for the Public Broadcasting Company VPRO and the WoZoCo housing for elderly in Amsterdam brought MVRDV to the attention of a wide field of clients and reached international acclaim.

Realized projects include the Dutch Pavilion for the World EXPO 2000 in Hannover, an innovative business park 'Flight Forum' in Eindhoven, the Silodam Housing complex in Amsterdam, the Matsudai Cultural Centre in Japan, Unterföhring office campus near Munich, the Lloyd Hotel in Amsterdam, an urban plan and housing in The Hague Ypenburg, the rooftop-housing extension Didden Village in Rotterdam, the cultural centre De Effenaar in Eindhoven, the boutique shopping building Gyre in Tokyo, Veldhoven's Maxima Medical Centre and the iconic Mirador housing in Madrid.

The grand variety of projects continues in the work of the office. Current projects in progress or on site include various housing projects in the Netherlands, Spain, China, France, Austria, the United Kingdom, USA and other countries, a television centre for Zürich, a public library in Spijkenisse (Netherlands), a central market hall in Rotterdam, a culture plaza in Nanjing, China, large scale urban masterplans in Oslo, Norway, Tirana, Albania and a masterplan for an eco-city in Logrono, Spain. Large scale visions for the future of greater Paris and the doubling in size of Dutch new town Almere are developed.

Naço Architecture

Add: 25 Jianguo Zhong Rd Bldg 6 Rm 9210, China
Tel: 86 21 6137 34 37; Fax: 86 21 6137 3431
Web: www.naco.net

Naço is a French Architects Studio based in Paris, Shanghai and Buenos Aires, known for its creative and innovative design solutions. The founder Marcelo Joulia set up the agency in Paris in 1991, now the dedicated global team includes 41 designers in architects, interior & graphic design.

In October 2005, Naço set up a design studio in Shanghai. The studio is located in Bridge 8 (phase 2) within the old French Concession, a well-known area with studios for creative designers, advertisers and fashion designers.

Naço Shanghai is a young and dynamic design agency with a total of 18 designers including architects, interior designers, product designers and graphic designers, mostly came from Europe. It's a real melting pot where different cultures brainstorm creative ideas.

Ong&Ong Pte Ltd.

Add: 510 Thomson Road, SLF Building #11-00,
Singapore 298135
Tel: +65 6258 8666; Fax: +65 6259 8648
Web: www.ong-ong.com

Founding partners, the late Mr. Ong Teng Cheong and Mrs.Ong Siew May, established Ong & Ong Architects in 1972. Since its humble beginnings, the firms' staff strength has grown to almost 500 over the past four decades. Going from strength to strength under robust and exceptional leadership, the organisation was incorporated in 1992.

With a track record of almost 40 years in the industry, Ong&Ong has earned an unparalleled reputation for integrating skilled architecture, clever interior design, creative environmental branding and sensitive landscape design. Paramount to their success lies in their insistence on servicing clients with creativity, excellence and commitment. They continually strive to uphold their mission to be the designer of this age – a premier design practice both locally and in the region.

Partnering their clients in their race to the top, Ong&Ong now offers a complete 360º solution – i.e. a parceled cross-discipline integrated solution, encompassing all aspects of the construction business. They offer a three-pronged 360º solution; namely 360º Design + 360º Engineering + 360º Management. 360º Design encompasses urban planning, architecture, landscape, interiors and environmental branding (graphics). Engineering 360º offers civil, structural, electrical, mechanical and plumbing. Management 360º provides development, project, construction and place management.

Pascal Arquitectos

Add: Atlaltunco #99, Col. lomas de Tecamachalco.
Cp 53970, Mexico D.F.
Tel: (5255) 52.94.23.71; Fax: (5255) 52.94.85.13
Web: www.pascalarquitectos.com

Pascal Arquitectos was founded in 1979 by Carlos and Gerard Pascal with the main purpose of achieving ultimate and integral development in architecture, interior, landscape, lighting and furniture design. Atelier's publication ranges from luxurious residential projects, residential complexes, institutional and religious buildings, restaurants, to corporate and offices buildings and hotels.

Patkau Architects

Add:1564 West 6th Ave, Vancouver BC,
Canada V6J 1R2
Tel: 604 683 7633, Fax: 604 683 7634
Web: www.patkau.ca

Patkau Architects is based in Vancouver, British Columbia, Canada. There are currently two principals: John and Patricia Patkau, and three associates: David Shone, Peter Suter and Gref Boothroyd.

In over 30 years of practice, both in Cananda and the United States, Patkau Architects has been responsible for the design of a wide variety of building types for a diverse range of clients. Projects vary in scale from galllery installations to master planning, from modest houses to majoy urban libraries. Their nany projects have involved functional programming, management of detailed public processes, and design of complex buildings and sites.

Their current work include a Visitors' Centre at Fort York Historic Site in Torontao. The Marcel A Desautel Faculty of Music and the School of Art at the University of Manitoba, the Golding Centre for High Performance Sport at the University of Toronto, a series of Cottages at Fallingwater, Mill Run, Pennsylvania, as well as a variety of residential projects diverse locations ranging from a northern island off the coast of British Columbia to a farm in Ad'Diriyyah, Saudi Arabia. As the circumstances of the work change, Patkau Architects' interest expand.

They seel to explore the full richness and diversity of architectural practice, understanding it as a critical act that engages their most fundamental desires and aspirations.

They refuse singular definitions of architecture: as art, as technology, as social service, as environmental agent, as political statement. They embrac all these definitions, togethet, as part of the rich, complex and vital discipline that they believe architecture to be.

Patrick Tighe Architecture

Add: 1632 Ocean Park Blvd / Santa Monica, CA 90405, USA
Tel: 310.450.8823
Web: www.tighearchitecture.com

Patrick Tighe is Principal and lead designer of Patrick Tighe Architecture. The firm is committed to creating an authentic, contemporary Architecture informed by technology, sustainability and building innovation. Since the inception, in 2001, a strong and diverse body of projects has been realized that include city developed affordable housing, commercial, mixed use projects, civic art, installations and residences.

In 2007, Patrick Tighe was awarded the prestigious Mercedes T. Bass Rome Prize in Architecture. The previous year, Tighe was the recipient of the American Institute of Architect's Young Architect Award and the 40 under 40 Award. Tighe is a Fellow of the American Academy and the MacDowell Colony. The work has received numerous awards including five National AIA Honor Awards, American Architecture Awards, a progressive Architecture Award, Los Angeles Architecture Awards, West Side Prize, Best of Year Awards as well as local AIA Honors.

Recently completed projects include the Sierra Bonita Mixed use Affordable Housing project for people living with disabilities. The 4645 aquare metres building serves as a pilot for the City of West Hollywood's newly implemented Green Building Ordinance. The Moving Picture Company (2009) is the US headquarters for the UK based post-production company located in Santa Monica.

Buildings have been realized in New York, Texas, Maine, Massachusetts and many of the culturally diverse communities within the greater Los Angeles area. Projects are now in development in Morocco, Asia and the Middle East. "Out of Memory"a site-specific installation designed by Patrick Tighe for the SCI-Arc Gallery, will be an experience at the convergence of sound, form, material, light and technology, the opening is February 4th 2011.

Projektil Architekti

Add: Malátova 13, Prague 5, 150 00, Czech Republic
Tel: +420 222 365 000; Fax: +420 222 365 014
Web: www.projektil.cz

The Projektil architekti studio was founded in 2002 by Roman Brychta, Adam Halíř, Petr Lešek, and Ondřej Hofmeister. The first completed project was Sluňákov – The Centre for Ecological Activities. This building won the Grand Prix of Czech Architecture in 2007. The following project which opened in September, 2008 was the regional Researche Library in Hradec Králové, which was awarded as the winner of the Grand Prix of Czech Architecture in 2009 in the "Best New Building"category.

The most recently completed project was the National Technical Library in Prague, which opened in September, 2009 and won the Grand Prix of Czech Architecture in 2010. The NTL was also shortlisted in the Great Indoors Award 2009. www.the-great-indoors.com

Projektil architekti – a studio of four young architects - designs private and public buildings, urban plans, interiors, and exhibition spaces. The architects are interested in innovations in typology and sustainable development. They invite experts, artists, scientists, and designers to collaborate on their projects. The Projektil architekti studio's motto is: teamwork, openness, and interdisciplinary discussion.

Richard Moreta Studio

Add: HQ Damaschkestr. 21 Erfurt 99096 Germany
Tel: (49) 162-214-7496 (Erfurt); Fax: (49) 162-214-7496
Web: www.gmz-design.com

Their works promote and explore design oriented solutions to contemporary urban issues, providing high quality, publically beneficial solutions to design, architecture and planning challenges.

Richard is the Designer and Project Manager Team Leader for development of hospitality, high end residential, and aviation, public, residential, industrial and commercial complex facilities at GMZ Design. He interacts with people from the contracting, project scheduling, financing and marketing sectors of the industry. Richard also works as a consultant to corporate clients based in Europe, Florida, and the Caribbean. For over the last 15 years, he has had an extensive background in Design and Construction, working in continental Europe, Africa, Asia, Caribbean, and the U.S.

Richard's work has been published in local and international magazines including, Abitare, Milan , Italy , Kenchi Bunka , Japan , HOME magazine, South Beach and others, he has also received an AIA Award of Excellence for "Vertropolis"in Saudi Arabia . Richard is a founding member GMZ, a group of recognized architectural designers with offices in New York, Miami and San Francisco.

ROBLESARQ Architecture Studio

Add: 11270-1000 San Jose, Costa Rica
Tel: 506 22802255, Fax: 506 2225 5418
Web: www.roblesarq.com

Roblesarq was founded in 2004 by architect Juan Robles. It is formed by a multidisciplinary team of young and creative professionals, distinguished by their aspiration to find innovative solutions. Roblesarq is recognized for its continuous research in all fields of architecrure, as well as for working with aesthetic art and technology.

Furthemore, it is also responsible for the integration of the natural environmental cycles in sustainable and bio-climatic "green"architecture. Its holistic approach, allows the exquisite use of space and structural order to blend with the adequate use of natural resources such as light, water and air flow. As a result, the atmospheres created not only seem to be sophisticated and dynamic contemporary ambiences, but true and unique sanctuaries of design, style and art.

Architect Juan Robles: LEED AP with specialty in Building design and construction, founder of RoblesArq in 2004 developing Residential, Commercial and Urban Planning based on sustainable architecture in different countries as Spain, Panamá and Costa Rica. He is an international business consultant in integral sustainable development based on the ecological principles (The Natural Step).

Rockwell Group

Add: 5 Union Square West, New York, New York 10003, USA
Tel: 212 463 0334; Fax: 212 463 0335
Web: http://www.rockwellgroup.com

With a desire to create immersive environments, Rockwell Group takes a cross-disciplinary approach to its inventive array of projects. Based in downtown New York with a satellite office in Madrid, their innovative, internationally acclaimed architecture and design firm specializes in hospitality, cultural, healthcare, educational, product, theater and film design.

Crafting a unique and individual narrative concept for each project is fundamental to Rockwell Group' successful design approach. From the big picture to the last detail, the story informs and drives the design. The seamless synergy of technology, craftsmanship and design is reflected in environments that combine high-end video technology, handmade objects, special effects and custom fixtures and furniture.

SAKO Architects

Add: 1803,1801 Tower8, JianWai Soho No.39, East Third Ring Road, Chaoyang District, Beijing 100022, China.
Tel: +86-10-5869-0901; Fax: +86-10-5869-1317
Web: www.sako.co.jp

Born in Fukuoka, Japan, as the principal of the firm, Keiichiro Sako was invited to work as visiting Scholar at Columbia University from 2004 to 2005. He established SAKO Architects and jointly presided over Asian Architects Associates in 2004. They created numerous works including JEANSWEST in Suzhou, ROMANTICISM 3 in Hangzhou, STEPS in Beijing, GOBI in Ulaanbaatar, EIFINI2 in Chengdu, KID'S REPUBLIC4 in Shanghai, FLATFLAT in Harajuku, EIFINI in Beijing, BRANCH in Changchun, MOSAIC in Beijing, LATTICE in Beijing, STRIPES in Jinan, BEANS in Kanazawa, T in Tokyo, and so on.

Heir works won many awards such as "JCD Design Award 2008 (Japan)", "Silver Award of Romanticism2 in Hangzhou", "Interior Design Award In China 2007", "The 5th Modern Decoration International Media Prize (China)", "Annual Best Designer", "Innovational top 10 on Tenment Design 2007 (China)", "Gold Award" of BUMPS in Beijing, "Science &Technology Award 2007 (China)", "Gold Award" of BUMPS in Beijing, "JCD Design Award 2007 (Japan)", "Bronze Award" of Romanticism in Hangzhou, "Asian Apartment Award 2006 (China)", and so on.

SB Architects

Add: San Francisco, California One Beach Street,
Suite 301 San Francisco, CA 94133 USA
Tel: (415) 673-8990
Web: www.sb-architects.com

In five decades of practice, SB Architects has established a world-wide reputation for excellence in the planning and design of large-scale hotel, resort, vacation ownership, multi-family residential and mixed-use projects. Widely recognized for their highly individual approach to design, SB Architects has received over 200 awards for design excellence. With a staff of highly trained, immensely talented and deeply passionate individuals in San Francisco and Miami offices, they successfully merge fifty years of experience with the energy, drive and dedication of a second generation of partners. With the technical capabilities and the understanding to take even the largest projects from initial concept through construction anywhere in the world, they remain a hands-on, design-oriented practice at heart. Integrity in design, connection to the client and balance in their corporate culture are integral to their identity.

Schmidhuber + Partner

Add: Nederlinger Strabe 21, D 80638, Munich, Germany
Tel: +49 89 15 79 97 -31
Web: www.schmidhuber.de

Schmidhuber+Partner are specialists in temporary buildings, corporate architecture und corporate design. For 25 years, the architects are successfully designing brand messages and corporate visions.

Today, over 50 architects, interior designers and designers work for international clients to realize trade shows and exhibitions, events and shops and showrooms. References are Audi, O2, Grohe, Lamborghini, Samsung and Stiebel Eltron. In 2010 they won 34 awards by ADC, CLIO, iF, ADAM & EVA, red dot and others. The German Pavilion at Expo 2010 in Shanghai was a cutting-edge project built on behalf of the Federal Ministry of Economics and Technology, which had commissioned Koelnmesse International to coordinate the preparations to run it.

The Consortium German Pavilion EXPO 2010 Shanghai was the general contractor on the project and thus responsible for planning and building balancity. Schmidhuber + Kaindl designed the pavilion architecture and was in charge of planning it. Milla & Partner from Stuttgart were responsible for designing the exhibition and multimedia features. NUSSLI, based in Roth near Nuremberg, managed the project and performed the construction work.

Simon Winstanley Architects

Add: 190 king street, castle douglas, dg7 1db, UK
Tel: 01556503826, Fax: 01556503828
Web: www.candwarch.co.uk

Simon Winstanley Architects was established in 1983. The practice specialise in contemporary architecture and energy conscious design having completed a wide range of projects including many award winning buildings. They can handle projects of all sizes and make an imaginative and innovative design response to every brief. Simon Winstanley Architects take pride in providing a high quality professional service and design with care and sensitivity to the character of the existing architecture, surrounding context and environment. They believe that a balanced approach is best which incorporates traditional building methods with contemporary ideas and that better standards deliver superior buildings which, together with considered and integrated landscaping, contribute to creating successful sustainable places.

Simon Winstanley Architects have developed strategies in their previous projects to allow Low Carbon Buildings to be procured at levels of low capital outlay. They can advise on systems considered appropriate. They can undertake Energy Audit's during the design process to determine the best solutions while also looking at the financial costs and returns on investment. Some features also qualify for grants and funding assistance. They aim to specify natural materials from appropriate sustainable sources, to reduce the overall carbon footprint of the building, both during its construction and together with the energy in use strategies described above, during its life span. Natural, durable external and internal finishes would be specified to minimize maintenance requirements.

Standard architecture

Add: 3521 dahlia avenue los angeles, california 90026 USA
Tel: 323 662 1000; Fax: 323 662 0199
Web: www.standard-la.com

Standard is the Los Angeles based architecture and design partnership of architects Jeffrey Allsbrook and Silvia Kuhle. Building on the partners international experience in a wide variety of project types, Standard is developing a large body of contemporary and diverse work. Completed projects include residential, retail, educational, cultural, office and manufacturing spaces for a diverse clientele of artists, writers, filmmakers, clothing designers, educators and entrepreneurs in California, New York, Las Vegas, Paris and Mexico. While Standard continues to grow, its partners insist upon maintaining a practice that is rigorous and attentive. Direct accessibility and sustained dialogue between clients and the firm partners and architects are viewed as essential to project success.

Studio Kalamar

Add: Studio Kalamar d.o.o. Slovenska 19,
1000 Ljubljana, Slovenia
Tel: +386 1 2410 470, Fax: +386 1 2410 473
Web: www.kalamar.si

While the majority of its work focuses on office and public buildings, the practice's projects range from masterplans for cities to interior and furniture design. In striving for design excellence in conjunction with technological innovation and functionality, the practice closely collaborates with clients and consultants from different fields. This active collaboration as well as the in-house supporting skills such as visualization, modelmaking and graphic design enable the practice to offer clients full support from the concept stages to final projects without delays.

From 1996 to 2010, the practice's completed projects have totalled over 250,000 square metres, all built on schedule and within their respective budgets. Andrej Kalamar: Born in 1966 in Murska Sobota, graduated from University of Ljubljana Faculty of Architecture, where he finished specialist studies in Urban Planning in 2003. Founded Studio Kalamar in 2003. His projects received several international architectural awards.

Terry G Green, Architect

Add: 5022 Sun Circle, Sarasota, Florida 34234, USA
Tel: 941 359-1815; Fax: 941 359-1815
Web: http://www.terryggreen.com

Back in 1969, when Terry Green was still cleaning up from the "muddy mess" he encountered at the Woodstock Music and Arts Fair, he was becoming interested in the ecology. Around that time, he was reading about counterculture lifestyle practices in the Whole Earth Catalog. And, while an architecture student at the University of Florida, he wrote his thesis on self-sustainable community design. So it follows, especially given his surname, that he has built a green house in Sarasota's Museum District. Built on speculation, the 334 square metres house recently won several awards for its design. He strives to create original high quality designs that honor their unique setting; that represent a sensitivity to environmental sustainability: and that reflect the individual character of the Owner's lifestyle.

Totan Kuzaembaev Architectural Workshop

Add: 105120, Nizhniaya Syromiatnicheskaya street, 10, build 3, office 34, Moscow, Russia.
Tel: +7-495-669-38-09, +7-495-514-17-89
Web: totan.ru

Located in Moscow, Rossia, Totan Kuzaembaev Architectural Workshop produce architectural projects in all stages, for all kinds of buildings.

TWS & Partners

Add: Wisma 21, Jl. Kembang Permai Raya Blok 1-4 No.3, Puri Indah, Kembangan Selatan, Jakarta 11610, Indonesia.
Tel: 62 21 582 8086; Fax:62 21 582 8427
Web: www.twspartners.com

Born on 30 January 1972, Jakarta, Indonesia, Tonny Wirawan Suriadjaja is an architect who always tries to find any innovation in architectural and design interior. His fields is ranging from urban scale to products, interior and furniture.

He always tries to discover something new that can improve his design and products. During 1990 and 1995, Tonny studied architecture at Tarumanegara University. In 1995 he worked with Mr. Gunawan Tiahjono Ph.d to design National Musuem in Korea. In the same year, he joined Shimizu Lampiri Consultant. Tonny was in Ciputra Development from 1995 to 1998. He is the principal at TWS & Partners since 1998.

TYIN tegnestue Architects

Add: Postbox 8848, 7481 Trondheim, Norway
Tel: +47 452 76 356
Web: www.tyintegnestue.no

TYIN tegnestue is a non-profit organisation working humanitarian through architecture. TYIN is run by Andreas Grøntvedt Gjertsen and Yashar Hanstad. Their projects are financed by more than 60 Norwegian companies, as well as private contributions. Through the course of the last year TYIN has worked with planning and constructing small scale projects in Thailand. They aim to build strategic projects that can improve the lives for people in difficult situations. Through extensive collaboration with locals, and mutual learning, TYIN hope that their projects can have an impact beyond the physical structures.

UNStudio

Add: PO Box 75381, 1070 AJ Amsterdam, The Netherlands
Tel: +31 (0)20 570 20 40; Fax: +31 (0)20 570 20 41
Web: www.unstudio.com

UNStudio, founded in 1988 by Ben van Berkel and Caroline Bos, is a Dutch architectural design studio specializing in architecture, urban development and infrastructural projects. The name, UNStudio, stands for United Studio referring to the collaborative nature of the practice.

At the basis of the practice are a number of long-term goals, which are intended to define and guide the quality of their performance in the architectural field. They strive to make a significant contribution to the discipline of architecture, continue to develop their qualities with respect to design, technology, knowledge and management and to be a specialist in public network projects.

They see as mutually sustaining the environment, market demands and client wishes that enable the work, and they aim for results in which their goals and their client's goals overlap. The office is composed of individuals from all over the world with backgrounds and technical training in

numerous fields. As a network practice, a highly flexible methodological approach has been developed which incorporates parametric designing and collaborations with leading specialists in other disciplines. Drawing on the knowledge found in related fields facilitates the exploration of comprehensive strategies which combine programmatic requirements, construction and movement studies into an integrated design.

Based in Amsterdam, the office has worked internationally since its inception and has produced a wide range of work ranging from public buildings, infrastructure, offices, residential, products, to urban masterplans.

VMDO Architects

Add: 200 E. Market Street Charlottesville, Virginia 22902
Tel: 434.296.5684; Fax: 434.296.4496
Web: http://vmdo.com

VMDO Architects, headquartered in Charlottesville, Virginia, has specialized in design for educational institutions since 1976. Their hallmark is the unwavering commitment they have to designing environments that shape the way people live, learn, work, and play, and at their best, uplift the human spirit.

Wilfrid Bellecour

Add: 4 boulevard de Strasbourg 75010 Paris, France
Tel : 01 40 40 07 07, Fax: 01 40 40 07 88
Web : www.studiobellecour.com

Wilfrid Bellecour was born the 11th January 1967. He achieved his B.A. from University of Houston. From 1994 to 2002, Wilfrid worked at Atelier Christian De Portzamparc. During that period, he participated in various projects all around the world. In 2003, he founded Studio Bellecour. Until Now, Studio Bellecour has dilivered many exellent projects including housings, offices, equipments and urban designs.

WHIM Architecture

Add: PO Box 11087, 3004 EB Rotterdam, the Netherlands
Tel: +31 (0)10 2623701
Web: www.whim.nl

The architecture firm WHIM architecture allows possibilities to create innovative design solutions generating an outstanding approach, that synthesizes the international experience of the Rotterdam based firm. The office was founded in 2005 by Ramon Knoester (NL).

The work is a combination of design and research. Committed to innovative and enjoyable architecture, WHIM seeks to respond to the unique circumstances and inherent potentialities of each project. Programmatic layout and sculptural form are the main assets of the office. Great care and attention is paid to the relations and dimensioning of program. Each building design results in a sculpture; a carefully studied form.

It is the opinion of the office that where society changes, architecture should adapt. A lot of time is paid to research, on how society changes and on how to react on it by architecture. WHIM made an award winning proposal on how to react on the growing amount of single households and a study on the ageing population. End of 2009, the office started a research project to make a new floating island, from the existing plastic waste that is floating in the Oceans. Recycled island is a proposal to clean the oceans from all the plastic pollution, create new land and set an example how to create a new sustainable habitat. This project can be followed on www.recycledisland.com.

WWAA

Add: Ul.Mińska 25, 03-808 Warszawa, Poland
Tel: +48 22 435 60 86. Fax: +48 22 435 60 87
Web: www.wwaa.pl

Architects Marcin Mostafa and Natalia Paszkowska are founding partners of WWAA – architectural practice which undertakes actions in many scales ranging from graphics, every-day use objects to furniture, urban objects, interiors, houses, office buildings as well as public buildings. They have been working together since 2003, with focus on projects for architectural contests, of which majority they won or were awarded prizes.

Marcin Mostafa was born in 1979 in Warsaw. MSc Diploma in Architecture from Warsaw University of Technology. In 2005 awarded the first prize and the execution of the project of a small pavilion in the competition organized in Warsaw. In 2007 a distinction in the prestigious competition for a design of an office building to be erected in the centre of Warsaw. Also involved in designing stage scenery for the theaters in Warsaw, Lodz and Radom, theatrical posters and clothes collections..

Natalia Paszkowska was born in 1981 in Katowice. MSc Diploma in Architecture from Warsaw University of Technology. Since 2004 has been acquiring professional experience in Poland's leading architectural offices. In 2005, together with Marcin Mostafa, awarded the first prize and the execution of the project of a small pavilion in the competition organized in Warsaw.

Boris Kudlička was born in 1972 in RuŁžomberok. A graduate of the Academy of Music in Bratislava 1996. He works regularly with several stage directors. Since 1999 Boris Kudlička has produced 12 operas with Mariusz Treliński. The operas were also presented at several opera festivals, including the Edinburgh International Festival, the Hong Kong Festival, and the Tokyo Festival. Currently, he is also cooperating with WWAA.

ZGF Architects LLP

Add: 1223 SW Washington Suite 200, Portland OR 97205, USA
Tel: 503-224-3860
Web: www.zgf.com

ZGF Architects LLP is an architectural, planning, and interior design firm with offices in Los Angeles, Portland, Seattle, Washington D.C., and New York. With a growing national design reputation, ZGF is involved in a diverse mix of both public and private projects nationwide, ranging from museums and educational buildings; to civic facilities; to regional airports and transportation systems; to commercial and corporate developments; to healthcare and research institutions, in settings ranging from campuses to urban centers.

They have been instrumental in the development of design guidelines and land use policies, programs, and master plans that have provided the framework for hundreds of millions of dollars in public and private investment.

ZGF's work has been recognized with more than 450 national, regional and local awards. In 1991, the firm received the industry's highest honor, the national Architecture Firm Award from the American Institute of Architects.

Their projects are featured in a number of publications, including seven monographs: Building Community, published by Rockport Publishers; Between Science and Art, published by L'Arca Edizioni; Building the Doernbecher Children's Hospital, Building the Conference Centre for the Church of Jesus Christ of Latter-day Saints, the Evolution of the Portland International Airport, and FDA at Irvine, all published by Edizioni Press; and, most recently, Future Tense published by Balcony Press.

图书在版编目（CIP）数据

住宅建筑 / 度本图书编著. -- 南京：江苏科学技术出版社，2014.3
（创新与平衡）
ISBN 978-7-5537-2808-7

Ⅰ. ①住… Ⅱ. ①度… Ⅲ. ①住宅－建筑设计－作品集－世界 Ⅳ. ①TU241

中国版本图书馆CIP数据核字(2014)第011462号

创新与平衡·住宅建筑

编　　　著	度本图书
项 目 策 划	凤凰空间
责 任 编 辑	刘屹立

出 版 发 行	凤凰出版传媒股份有限公司 江苏科学技术出版社
出版社地址	南京市湖南路1号A楼，邮编：210009
出版社网址	http://www.pspress.cn
总 经 销	天津凤凰空间文化传媒有限公司
总经销网址	http://www.ifengspace.cn
经　　　销	全国新华书店
印　　　刷	北京博海升彩色印刷有限公司

开　　　本	965 mm×1 270 mm　1/16
印　　　张	20
字　　　数	160 000
版　　　次	2014年3月第1版
印　　　次	2014年3月第1次印刷

标 准 书 号	ISBN 978-7-5537-2808-7
定　　　价	248.00元

图书如有印装质量问题，可随时向销售部调换（电话：022-87893668）。